Spatial-Temporal Evolution of Mining-Induced Rock Damage and Ground Control of Roadways

Zhijie Wen · Zhenqi Song · Yujing Jiang ·
Yujun Zuo · Jian Tao

Spatial-Temporal Evolution of Mining-Induced Rock Damage and Ground Control of Roadways

Zhijie Wen
Mining College
Guizhou University
Guiyang, Guizhou, China

Yujing Jiang
Graduate School of Engineering
Nagasaki University
Nagasaki, Japan

Jian Tao
Mining College
Guizhou University
Guiyang, Guizhou, China

Zhenqi Song
College of Energy and Mining Engineering
Shandong University of Science
and Technology
Qingdao, Shandong, China

Yujun Zuo
Mining College
Guizhou University
Guiyang, Guizhou, China

ISBN 978-981-96-5438-3   ISBN 978-981-96-5439-0 (eBook)
https://doi.org/10.1007/978-981-96-5439-0

This work was supported by Shandong Province College Youth Talent Introduction and Cultivation Support Program, ZR2019YQ26 (Shandong Province Outstanding Youth Fund) and No. 51974174 (National Natural Science Foundation of China).

© The Editor(s) (if applicable) and The Author(s) 2025. This book is an open access publication.

Open Access This book is licensed under the terms of the Creative Commons Attribution-NonCommercial-NoDerivatives 4.0 International License (http://creativecommons.org/licenses/by-nc-nd/4.0/), which permits any noncommercial use, sharing, distribution and reproduction in any medium or format, as long as you give appropriate credit to the original author(s) and the source, provide a link to the Creative Commons license and indicate if you modified the licensed material. You do not have permission under this license to share adapted material derived from this book or parts of it.

The images or other third party material in this book are included in the book's Creative Commons license, unless indicated otherwise in a credit line to the material. If material is not included in the book's Creative Commons license and your intended use is not permitted by statutory regulation or exceeds the permitted use, you will need to obtain permission directly from the copyright holder.

This work is subject to copyright. All commercial rights are reserved by the author(s), whether the whole or part of the material is concerned, specifically the rights of translation, reprinting, reuse of illustrations, recitation, broadcasting, reproduction on microfilms or in any other physical way, and transmission or information storage and retrieval, electronic adaptation, computer software, or by similar or dissimilar methodology now known or hereafter developed. Regarding these commercial rights a non-exclusive license has been granted to the publisher.

The use of general descriptive names, registered names, trademarks, service marks, etc. in this publication does not imply, even in the absence of a specific statement, that such names are exempt from the relevant protective laws and regulations and therefore free for general use.

The publisher, the authors and the editors are safe to assume that the advice and information in this book are believed to be true and accurate at the date of publication. Neither the publisher nor the authors or the editors give a warranty, expressed or implied, with respect to the material contained herein or for any errors or omissions that may have been made. The publisher remains neutral with regard to jurisdictional claims in published maps and institutional affiliations.

This Springer imprint is published by the registered company Springer Nature Singapore Pte Ltd.
The registered company address is: 152 Beach Road, #21-01/04 Gateway East, Singapore 189721, Singapore

If disposing of this product, please recycle the paper.

Preface I

Deep coal resources are the reserve of the main energy source in our country in the twenty-first century. If old mining areas are closed due to the exhaustion of shallow resources, the energy in the developed eastern regions will become even more tense. Therefore, the safe development of deep coal resources, ensuring the continued production of eastern mining areas, and providing sufficient coal supply for national economic and social development will be beneficial to ensure the country's energy security.

This book aims to initially reveal the occurrence and effective control of major disasters in mining areas, and proposes a framework for predicting and controlling major accidents by controlling the movement and stress conditions of mining surrounding rocks. It provides a reference basis for promoting related research. The main content of the manuscript is the author's summary of the research on mining dynamics and rock layer control in the past 10 years. The main content has received the careful guidance and strong support of Academician Zhenqi Song, to whom I would like to express my sincere thanks and high respect! During the writing of the manuscript, I also received the careful guidance of Prof. Yongkui Shi, Assoc. Prof. Guozhi Lu, Prof. Chongge Wang, and other teachers from the Academician team, and quoted some research results, for which I express my heartfelt thanks!

Thanks to Renle Zhao, Chief Engineer of Shandong Energy Lin Mine Group, Chunfeng Liu, Director, and Yangyang Li, Lishuai Jiang, Guangchao Zhang, Nasser, Hengjie Luan, Hengzhong Zhu, and others from the Mining Dynamics Research Group for their support during the writing of the manuscript.

The publication of this book was supported by the National Natural Science Foundation of China (52274130), National Key R&D Program (2016YFC0600708), Shandong Provincial Natural Science Foundation (ZR2018MEE001), Taishan Scholars Advantage and Characteristic Discipline Talent Team Support Plan, Shandong Provincial Taishan Scholars Project Funding Support, Shandong Provincial Higher

Education Research Plan Project (Science and Technology) Key Project (J18K2010), Qingdao Source Innovation Plan (18-2-2-68-jch), National Key Laboratory Open Fund (SHGF-18-13-30), etc.

Due to the limited level of the author, there are inevitably omissions and deficiencies in the book, and I sincerely ask seniors and colleagues to teach.

Guiyang, China   Zhijie Wen
February 2025

Preface II

Coal is the main energy source in our country, with a total amount of coal resources less than 2000 m deep being 5.9 trillion tons, of which more than 50% are deeper than 1000 m, mainly distributed in the central and eastern regions of our country. Deep mining will become the new norm for the development of the coal industry and resource exploitation. To ensure the energy supply for the rapid economic development in the central and eastern regions, the development of kilometer-deep coal resources is inevitable, which is of great strategic significance for ensuring national energy security and supporting local economic development.

Accidents such as impact ground pressure, water seepage, and roof accidents caused by mining are still frequent, occurring from time to time in local coal mines and state-owned coal mines. This situation seriously threatens the safety of coal mining in our country and affects the image of our mining industry. Therefore, it is necessary to further improve mining engineering, especially the theory of predicting and controlling major accidents, to deeply interpret the dynamic information basis of mine pressure and rock layer movement related to the occurrence of accidents, to advance the decision-making and implementation management of safe and efficient coal mining to the stage of scientific quantitative development, and to realize informatization, intelligence, and visualization. This is also an urgent task to fundamentally solve the current safety situation of our mines.

In recent years, the author and his team have conducted in-depth research on the disaster-causing model of deep mine dynamic disasters and surrounding rock control under the support of the National 973 Program, the National Key R&D Program, talent and surface projects, and the Chinese Academy of Sciences consulting special projects. The main contents include the construction of the overburden spatial structure model of the mining area, the spatiotemporal evolution of mining stress, and the pre-control technology of dynamic disasters in the mining area.

The publication of *Spatial-Temporal Evolution of Mining-Induced Rock Damage and Ground Control of Roadways* will promote the in-depth development of research on mine pressure and rock layer control, and make certain contributions in talent cultivation, control of deep mine dynamic disasters, and promotion of technological progress.

Qingdao, China   Zhenqi Song

Competing Interests The authors have no competing interests to declare that are relevant to the content of this manuscript.

Contents

1 **Introduction** 1
  1.1 Mining Dynamics and Surrounding Rock Control 1
  1.2 Spatiotemporal Evolution Effect of Overburden Spatial Structure in Mining Area 4
    1.2.1 Analysis of Dynamic Evolution Characteristics of Overburden Spatial Structure in Mining Area in Full Time Domain 4
    1.2.2 Study on the Aging of Coal Body Damage and Fracture ... 6
  1.3 The "Time–Space" Disaster Incubation Process of Mining Stress Field 8
  References 9

2 **Research on Spatio-Temporal Migration Law of Overburden Structure in Mining Area** 13
  2.1 Construction of Overburden Spatial Structure in Mining Area 13
    2.1.1 Overview of the Three-Dimensional Spatial Structure Model of the Mining Area 14
    2.1.2 Classification of Overburden Spatial Structure in Mining Area 28
  2.2 The Evolution and Development Law of Overburden and Mining Stress in the Mining Field 31
    2.2.1 The Evolution and Development Process of Overburden and Mining Stress in the Mining Field 31
    2.2.2 Distribution Pattern of Mining Stress on the Plane 39
    2.2.3 Relationship Between Overburden Failure Structure and Mining Stress Evolution 46
  References 58

3 Temporal and Spatial Evolution Mechanism of Mining Stress Field 61
- 3.1 Definition of Deep Coal Mining 61
  - 3.1.1 Deep Coal Mining 62
  - 3.1.2 Deep Judgment Criteria and Determination Methods 63
- 3.2 Research on Coal Body Damage Evolution Characteristics 67
  - 3.2.1 Analysis of Coal Rock Damage Mechanism 67
  - 3.2.2 Characteristics of Coal Rock Damage Evolution 73
- 3.3 Study on the Spatiotemporal Evolution Law of Mining Stress Field 94
  - 3.3.1 Mining Stress Distribution Characteristics 94
  - 3.3.2 Study on the Spatiotemporal Characteristics of Mining Stress Evolution 98
- References 121

4 Key Technologies for Controlling Mining Dynamic Disasters 125
- 4.1 Key Technologies for Rock Burst Disaster Prevention and Control 126
  - 4.1.1 Overview of Rock Burst Disasters 127
  - 4.1.2 Rock Burst Occurrence Mechanism 141
  - 4.1.3 Basic Dynamic Information for the Prevention and Control of Impact Ground Pressure Disaster Accidents 145
- 4.2 Key Technologies for Pre-control of Roof Water Inrush Disaster 157
  - 4.2.1 Overview of Roof Water Inrush Disaster 157
  - 4.2.2 Mechanism of Roof Seepage Disaster 164
  - 4.2.3 Dynamic Information Basis for the Prevention and Control of Roof Seepage Disasters 165
- 4.3 Key Technologies for Roof Disaster Prevention and Control 174
  - 4.3.1 Overview of Roof Disaster Accidents in the Mining Face 174
  - 4.3.2 Mechanism of Roof Disaster Accident 177
  - 4.3.3 Basic Information for Roof Disaster Accident Prevention and Control 179
- References 197

5 Control of Large Deformation of Surrounding Rock Based on Stress Gradient Theory 199
- 5.1 Phenomenon of Large Deformation and Destruction of Surrounding Rock in the Mining Area 201
  - 5.1.1 Characteristics and Causes of Large Deformation and Damage of Surrounding Rock in Mining Area 201
  - 5.1.2 Mechanical Characteristics of Large Deformation of Stope Surrounding Rock 206

5.2	\multicolumn{2}{l}{Research on Rock Deterioration Model Based on Stress Gradient Theory}	213

5.2 Research on Rock Deterioration Model Based on Stress Gradient Theory 213
    5.2.1 Rock Mass Deterioration Parameters Based on Stress Gradient Theory 214
    5.2.2 Solution of the Rock Mass Deterioration Model Based on Stress Gradient Theory 216
5.3 Determination of Reasonable Pre-tightening Force of Anchor Bolt Based on Stress Gradient 224
    5.3.1 Determination of Reasonable Pre-tightening Force of Anchor Bolt by Theoretical Calculation 224
    5.3.2 Determination of Reasonable Pre-tightening Force of Anchor Bolt by Numerical Simulation 225
    5.3.3 Analysis of Theoretical and Measured Correlation 228
References 233

Chapter 1
Introduction

1.1 Mining Dynamics and Surrounding Rock Control

Coal is the pillar energy of our country, it is an important non-renewable fuel and industrial raw material, and the coal industry is an important basic industry related to the economic lifeline and energy security of our country. For a long time, coal has dominated the production and consumption structure of primary energy in our country [1–3]. In 2018, China's raw coal output was 3.58 billion tons [2], accounting for 59% of the energy consumption structure. It is estimated that by 2030, the proportion of coal will still be as high as about 50% [3]. Deep coal resources are the backup reserves of China's main energy in the twenty-first century. If old mining areas are closed due to the exhaustion of shallow resources, the energy in the developed areas of the east will become more tense. Therefore, the development of deep coal resources, ensuring the continued production of eastern mining areas, and providing sufficient coal supply for national economic and social development, is an important guarantee for national energy security.

However, as the mining depth of coal mines in China continues to increase, the impact of dynamic disasters is becoming more serious (Fig. 1.1), and the number of mines at risk of impact will continue to increase [4–7]. At present, more than 150 pairs of impact ground pressure accidents have occurred, spread across the main coal mining areas in China [8]; at the same time, related dynamic disasters have caused serious casualties and property losses: in 2015, the "7.29" impact ground pressure accident at Zhaolou Coal Mine of Shandong Yankuang Group caused 5 injuries, and in 2014, the "3.27" impact ground pressure accident at Qianqiu Coal Mine of Henan Yimei Group caused 6 deaths. According to incomplete statistics, China has accumulated more than 31,000 dynamic disasters, with an average of nearly 300 deaths per year [9], seriously affecting the international image of China's coal industry. At present, except for a few provinces and regions such as Hainan, Guangdong, Fujian, Zhejiang, and Tibet, the main coal mining provinces in China are threatened by dynamic disasters to varying degrees. The main mines in the famous

Fig. 1.1 Damage to the roadway after the disaster

Pingdingshan, Huainan, and Yanzhou mining areas have all become prominent mines. Therefore, the monitoring, prediction, and control of coal mine dynamic disasters have become the key issues for the healthy development of China's coal industry. China has clearly identified "mine gas, water inrush, dynamic disaster early warning and control technology" as a priority research topic in the "National Medium and Long-term Science and Technology Development Plan (2006–2020)" [10].

The essence of dynamic disaster phenomena such as impact ground pressure is the sudden instability and destruction of coal and rock bodies under high stress conditions. Compared with other underground engineering industries such as underground factories, hydroelectric caverns, and subway tunnels, coal mining has very distinct characteristics [11] (Fig. 1.2): The mining space is large, China's deep coal mines generally use longwall mining methods, forming hundreds of thousands or even millions of cubic meters of mining space. The large scale of mining and the wide range of mining disturbances are unmatched by any other underground engineering; the mining disturbance is strong, the longwall mining method of large space rapid mining forms a strong mining disturbance to the surrounding rock, causing the overlying rock layer to collapse and the ground surface to deform and sink on a large scale. Especially for the tens of millions of tons of mines that are mined on one side in the deep part, the large-scale strong disturbance caused by mining is unmatched by shallow mining and other underground engineering [12, 13]; The properties of the medium and the stress state are complex. In addition to the complexity of the geological conditions in deep coal fields, large-scale mining repeatedly disturbs the coal and rock mass around the mining space, causing it to undergo deformation and destruction multiple times. This results in the medium properties of the coal and rock mass having both discontinuous structural characteristics and broken medium

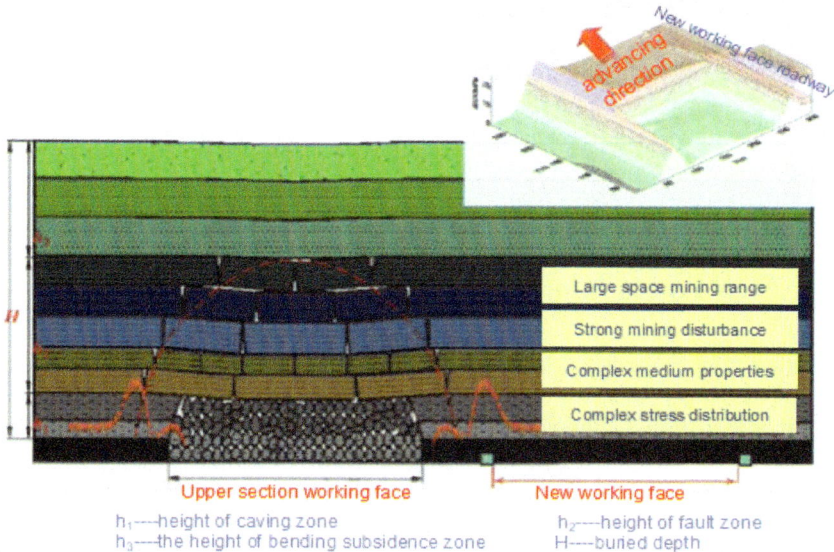

Fig. 1.2 Mechanical structure of the deep mining area

properties. The working face is under the combined action of high ground stress and strong unloading. The spatiotemporal relationship of stress redistribution induced by mining is complex. The dynamic characteristics of energy dissipation and energy release processes caused by high stress release, transfer, and transmission are obvious, which can easily induce dynamic disasters such as rock bursts.

The "mining-induced stress" that prompts the surrounding rock to move towards the mined space is the root cause of major accidents such as coal mine roof, gas, rock burst, and subsidence disasters. Not understanding or not fully mastering the spatiotemporal variation rules of rock layer movement and mining-induced stress distribution under different mining conditions, and excavating, maintaining roadways, and advancing the working face at the wrong time and space is one of the important reasons for dynamic disasters such as coal and gas outbursts and rock bursts.

In view of this, this book establishes a structural mechanics model for disaster-causing conditions in deep mining areas, divides the overburden spatial structure of the mining area into the moving rock layer structure "fracture arch" that directly affects the mine pressure manifestation of the mining area and the rock layer structure "stress arch" that has not produced obvious movement, and studies their correlation; based on acoustic emission theory and strain energy theory, it conducts a detailed analysis of the damage constitutive relationship of coal and rock mass, studies the spatiotemporal evolution mechanism of mining-induced stress field; proposes an evaluation mechanism for mining-induced stress breeding and unloading energy release based on energy dissipation rate index for impact dynamic disasters, a "single key layer" and "double key layer" structure's roof control rules for roof disasters, and a large deformation study of roadway based on the stress gradient theory of continuous damage mechanics model. The results of this book can provide theoretical support for studying the disaster-causing conditions of deep dynamic disasters, and at the same time make new contributions to dynamically controlling the disaster-causing environment, reducing or changing the disaster-causing conditions.

1.2 Spatiotemporal Evolution Effect of Overburden Spatial Structure in Mining Area

Before mining, the coal body is under a three-dimensional stress balance in the deep. Mining activities break the original stress balance, causing the macro stress field and energy field in the three-dimensional space of the mining area to redistribute. The dynamic evolution and development of this stress field and energy field create conditions for the breeding, occurrence, and development of dynamic disasters. The relationship between the fracture morphology and stress field of the space around the coal mine mining area is the basis for predicting and controlling dynamic disasters such as rock bursts, mine water inrush, coal and gas outbursts, and overall roof fall. Therefore, combining the two issues of "spatiotemporal effect of overburden spatial structure in coal mine mining area" and "spatiotemporal evolution characteristics and disaster-causing mechanism of mining-induced stress field", a comprehensive study of dynamic disasters such as rock bursts faced by deep coal mining can effectively reveal the mining effect of coal and rock dynamic disasters, which is conducive to the effective prevention and control of dynamic disasters such as rock bursts.

For a long time, many beneficial research works have been carried out on the full-time domain stability of the overburden spatial structure in the mining area, the damage and fracture characteristics of the coal and rock mass in the mining area, and the evolution law of the mining-induced stress field in deep mining areas [14–87].

1.2.1 Analysis of Dynamic Evolution Characteristics of Overburden Spatial Structure in Mining Area in Full Time Domain

Regarding the research on the damage range and structural characteristics of the overburden in the mining area, experts and scholars in the mining engineering field at home and abroad have carried out a lot of research work from the needs of explaining the manifestation rules of mine pressure in the mining area, solving the control and support design of the roof in the mining area, and the research on "three-under" coal mining.

1. Spatiotemporal Evolution Theory of Overburden Spatial Structure in Mining Area

Academician Qian and others [14–18] The theories of "masonry beam" and "key layer" have been established, providing a theoretical basis for the study of the formation and instability of overburden structures in coal mines; Academician Song and others [19] proposed the "transfer rock beam" theory, providing theoretical guidance for the study of roof control design; Jiang and others [20] based on the "masonry beam" and "transfer rock beam", proposed three basic structures of the basic roof: arch, arch beam, and beam; Shi and Deng [21] borrowed the mechanical analysis method of arch shell structure to make a preliminary analysis of the arch structure characteristics of the overlying rock layer movement of the caving coal mining field from a macro perspective; Yan and Jia [22] analyzed the reasons for the transfer of the balance structure of the overlying rock layer of the caving coal mining to a high position; Zhang [23] proposed the basic form of the overburden structure of the comprehensive mining face composed of "masonry beam" and "semi-arch" structure; Huang [24] established the "short masonry beam" and "step rock beam" structure model of the basic roof periodic pressure of the shallow buried coal mining field, these studies and their main results have laid the foundation for the study of the spatial structure theory of the overburden of the mining field. The overburden of the "fracture zone" of the mining field undergoes a "bending-fracture-touching the gangue-compacting the gangue" process, making the spatial structure of the overburden of the mining field evolve with the development of mining, and the existing research results mainly target the abstract two-dimensional structure mechanics model under static, shallow-middle deep conditions; therefore, the "four-dimensional" spatial structure full-time domain evolution model of the overburden of the mining field containing the mining time factor in the deep mining process still needs to be further studied and explored [25–28].

2. Monitoring of the spatiotemporal evolution process of overburden spatial structure in the mining field.

Scholars at home and abroad have done a lot of practical work to study the dynamic evolution of the overburden spatial structure in the mining field. At present, there are mainly four monitoring research methods.

The microseismic positioning monitoring method of rock fracture is used to monitor the formation process and approximate range of the overburden spatial structure. Microseismic Monitoring Techniques, abbreviated as MS technology [29–44], has been used for over a decade to monitor rock fractures on the scale of the working face. It is mainly used in shallow coal seams abroad, where detectors are installed by drilling from the surface. In China, due to the deep mining of coal mines, surface monitoring is not economically efficient or reliable. Therefore, domestic scholars have developed a deep well explosion-proof MS system and related deep hole detector installation technology based on mining conditions, to monitor the dynamic fracture process of the rock mass in three-dimensional space and determine the range and degree of fracture. Based on microseismic positioning monitoring technology, Jiang [45] divided the overlying rock space structure of the mining area according to the boundary conditions of the working face, and proved through the analysis of the measured data of six longwall faces in Australian coal mines that before the strata fully entered the mining, the maximum fracture height of the overlying rock layer is approximately half the short side length of the mined-out area.

The double-end water plugging leakage observation method [46, 47]—used to determine the height and degree of rock fracture in one direction at a point. The borehole stress meter method [48, 49]—Used for the inversion of overburden movement conditions and monitoring the stability of coal bodies. Numerical calculations and similar material simulation experiments [50, 51]—Used as an auxiliary means to study the spatial structure and stress field of overburden. Numerical calculations are used to study the stress field in the surrounding rock of the established overburden spatial structure, while three-dimensional similar material simulation experiments are mainly used for conceptual research of overburden spatial structure.

The evolution of overburden failure is not only related to the rock layer structure, but also has a strong time effect [52]. In fact, the problem of rock layer control is closely related to the time factor. Therefore, based on the research results of scholars at home and abroad, the time effect of mining is fully considered, and the spatiotemporal evolution law of the formation to stability of the overburden spatial structure in the mining area is studied and analyzed.

1.2.2 Study on the Aging of Coal Body Damage and Fracture

The mechanical environment, basic mechanical behavior, and failure characteristics of deep coal mining will gradually intensify with the deterioration of coal resource conditions, especially the coal body is disturbed by mining in the "formation-stability" full-time domain process of the overburden spatial structure in the deep mining area. Correct understanding and description of the evolution of coal body damage and fracture are of great practical significance for a deep understanding of the occurrence mechanism of dynamic disasters in deep mining areas.

Energy conversion is the essential characteristic of the physical process of matter, and material destruction is a state instability phenomenon driven by energy. The study of the physical process of coal body damage and fracture, especially considering the non-uniform evolution process of the deformation field in the process of coal body damage and fracture and the release and transfer of energy, has always been a hot issue in the field of rock science. Scholars at home and abroad have carried out a lot of research on this [53–55]. Obert [56] first applied acoustic emission monitoring technology to predict the instability phenomenon of rock excavation, and determined the maximum stress area in the rock by this; Cao [57] and others established a statistical damage constitutive model reflecting the whole process of rock fracture based on the rock uniaxial stress–strain curve; Zhu, Feng and others [58, 59] based on the scanning electron microscope (SEM) rock fracture full process digital micro-damage mechanics test scheme, realized the rock mechanics test of real-time digital monitoring, control, recording and analysis of the micro and macro of the rock fracture full process, described the failure mechanism of the rock sample during the uniaxial compression process from the macro and micro perspectives,

1.2 Spatiotemporal Evolution Effect of Overburden Spatial Structure …

and analyzed that although micro-cracks are concentrated in some areas during the uniaxial compression failure process of the sample, the statistical distribution of micro-cracks in the entire sample still follows a certain exponential distribution; Tang and others [60–63] The evolution process and interaction mechanism of rock cracks under various conditions were studied using numerical simulation; Liu and Liu [64] Based on the principle of minimum energy consumption, the intrinsic relationship between stress, energy accumulation transfer, and spatial distribution of microseismic activity was revealed; Zhao [65] The study found that the three-dimensional positioning results of acoustic emission events intuitively reflect the initial position, expansion direction, width change, evolution process of rock sample cracks, and the surface morphology of crack expansion, and can also reflect the evolution of the internal stress field of the rock; Cao and others [66] Studied the trend of AE signal changes of coal samples under different confining pressures, and compared it with the uniaxial situation; Zhao and others [67] Studied the acoustic emission characteristics of gas-bearing coal samples during triaxial compression, and established a damage equation for gas-bearing coal based on acoustic emission characteristics; Zuo and others [68] Conducted uniaxial and triaxial compression tests on Qianjiaying rock mass, coal body and coal-rock combination, and obtained the failure modes and mechanical behaviors of coal and rock, monomer and combination under different stress conditions; Shkuratnik and others [69–71] Conducted cyclic loading tests on coal bodies in the laboratory, and studied the memory effect of acoustic emission of coal and rock under complex stress; Voznesenskii and others [72] Studied the AE parameter characteristics of the upper and lower parts of the specimen during the loading process, and focused on analyzing and comparing their differences in the main crack formation stage. The above research results have an important role in promoting the understanding and quantification of the dynamic evolution process of coal body damage and fracture. However, compared with rocks, there are fewer research results on the acoustic emission characteristics of the load deformation and failure process of large deformation coal and rock, mainly for four reasons: first, coal and rock have obvious heterogeneity and anisotropy, and the typical elastic–plastic rock material AE characteristics cannot simply describe the properties of coal and rock; second, from the perspective of experimental implementation, the requirements for instrument precision and noise resistance are high; third, the loading method should realistically simulate the entire process of coal body subjected to mining field overburden fracture impact and slow dynamic loading after touching gangue; fourth, the experimental scale should reasonably avoid the scale effect caused by small specimen size.

The mechanical environment of deep coal mining is complex, and the test equipment should realistically simulate the conditions of coal body subjected to the impact and loading conditions of overlying rock fracture in the mining field. At the same time, the study of coal body acoustic emission mainly focuses on medium and low confining pressure (0–10 MPa). Therefore, further research is needed on the spatial evolution characteristics of AE time series under high confining pressure and the spatial evolution law of deep coal body damage and fracture under dynamic stress under the action of reasonable loading criteria.

1.3 The "Time–Space" Disaster Incubation Process of Mining Stress Field

People began to systematically study the phenomenon of deep dynamic disasters in the 1950s. Scholars in our country have also done a lot of research on the mechanism of dynamic disaster occurrence, field monitoring and prediction methods, and prevention and control technologies [73–77], but at present, dynamic disasters are still difficult to control, and the effects of prediction and prevention need to be further improved.

Jiang and others [78–81] used the stress drop value and stress change range to monitor and warn of seismic-induced rock burst in real time, predicting the key layer fracture step before mining, real-time dynamic analysis of rock layer fracture height during mining, judging the area of seismic occurrence, improving the timeliness and accuracy of on-site warning, and believed that the time interval between stress drop and seismic events varies from 3 to 8 min; Dou [82–84] based on the theory of key layer movement of rock layer, studied the fracture law of overlying rock key layer, analyzed the mechanical mechanism of "O-X" type fracture of overlying rock key layer, when the mine pressure reaches the maximum value, it is very likely to cause strong mine seismic and rock burst; Xie [85, 86] proposed that there is a macro stress shell composed of high stress bundles in the surrounding rock of the comprehensive mining face, only when the stress shell is unbalanced will it cause severe mine pressure phenomena, such as mine seismic, rock burst, etc.; Qi [25] It is believed that the occurrence of coal mine impact ground pressure usually occurs within the range of mining stress influence in front of the working face, under the influence of stress concentration and mining, leading to the occurrence of impact ground pressure; Yu [87] believes that the occurrence of impact ground pressure is closely related to the roof pressure, generally occurring 1–3 days after the strong periodic pressure of the working face.

These research results provide a certain theoretical basis and technical guidance for the study of the breeding mechanism and quantitative evaluation method of mining stress field in the whole time domain movement process of overburden spatial structure in deep mining area. However, the spatio-temporal effect of coal body damage degradation in the dynamic evolution process of mining area space and the dynamic evaluation method of mining stress field breeding caused by coal body damage have not been fully understood, which is mainly caused by the high complexity of the evolution of overburden spatial structure in mining area and the uncertainty of mining stress. Therefore, the formation of a stable full-time domain process of overburden spatial structure in deep mining area, the evolution process of mining stress field and the quantitative evaluation method aiming at dynamically controlling the breeding environment of dynamic disasters and reducing or eliminating the disaster conditions of dynamic disasters still need further research.

References

1. Xie HP, Wu LX, Zheng DZ (2019) Prediction of China's energy consumption and coal demand in 2025. J Coal 44(07):1949–1960
2. Qian MG, Xu JL, Wang JC (2018) Re-discussion on the scientific mining of coal. J Coal 43(01):1–13
3. Wu Q, Tu K, Zeng YF et al (2019) Discussion on the main problems and countermeasures faced by the upgrading of China's main energy (coal). J Coal 44(06):1625–1636
4. Song ZQ, Jiang YJ, Yang ZF et al (2003) Research on the dynamic information basis for the prediction and control of major accidents in coal mines. Coal Industry Publishing House, Beijing
5. Tan YL, Guo WY, Xin HQ et al (2019) Research on key technologies for monitoring and resolving dangers of impact ground pressure in deep mining of coal mines. J Coal 44(01):160–172
6. He MC, Xie HP, Peng SP et al (2005) Research on rock mechanics in deep mining. J Rock Mech Eng 24(16):2803–2813
7. Wen ZJ, Jing SL, Song ZQ et al (2019) Research on mining field spatial structure model and related dynamic disaster control. Coal Sci Technol 47(1):52–61
8. Pan JF, Mao DB, Lan H et al (2013) Current status and prospect of research on impact ground pressure prevention technology in China's coal mines. Coal Sci Technol 41(06):21–25
9. He XQ (2011) Progress in monitoring and early warning technology for coal and rock dynamic disasters in coal mines. Sci Times (A03)
10. (2006) National medium and long-term science and technology development plan outline (2006–2020)
11. Pan YS, Li ZH, Zhang MT (2003) Research on the distribution, type, mechanism and prevention of impact ground pressure in China. J Rock Mech Eng 22(11):1844–1851
12. Zhang JM, Li QS, Zhang Y et al (2019) Definition and mining response analysis of deep coal mining. J Coal 44(5):1314–1325
13. Qi Q, Li Y, Zhao S et al (2019) 70 years of development of coal mine impact pressure in China: establishment and reflection on the theoretical and technical system. Coal Sci Technol 47(09):1–40
14. Qian M, Miao X, Xu J (1996) Research on the key layer theory in strata control. J Coal 21(03):225–230
15. Xu J, Qian M (2000) Method for determining the position of the key layer in overburden. J China Univ Min Technol 29(05):463–467
16. Miao X, Qian M (2000) New progress in the study of the key layer theory of mining rock mass. J China Univ Min Technol 29(01):25–29
17. Qian M, Miao X, He F (1994) Analysis of the key block of the "masonry beam" structure in the mining area. J Coal 19(06):557–563
18. Qian M, Miao X, Xu J et al (2003) Theory of key layer in strata control. China University of Mining and Technology Press, Xuzhou
19. Song Z (1988) Practical mine pressure and control. China University of Mining and Technology Press, Xuzhou
20. Jiang F (1995) Design of mine roof control and its expert system. China University of Mining and Technology Press, Xuzhou
21. Deng G (1994) Study on the movement and destruction law of overlying strata in caving coal mining. Mine Pressure Roof Manag (02):23–26
22. Yan S, Jia G, Liu X (1996) Analysis of the mechanism of overlying strata structure transfer to high position in caving coal mining. Mine Pressure Roof Manag (03):3–5
23. Zhang D, Wang Y (1998) Analysis of strata structure in fully mechanized caving coal working face. J China Univ Min Technol (04):340–343
24. Huang Q (2000) Study on roof structure and strata control in longwall mining of shallow coal seam. China University of Mining and Technology Press, Xuzhou

25. Qi Q, Chen S, Wang H et al (2003) Relationship between impact pressure, rock burst, mine earthquake and their numerical simulation study. Chin J Rock Mech Eng 22(11):1852–1858
26. He M, Jiang Y, Zhao Y (2005) Impact pressure control theory centered on composite energy conversion. In: Fundamental theory research and engineering practice of deep resource mining. Science Press, Beijing
27. Lu CP, Dou LM, Wu XR (2006) Weakening control mechanism and its practice of coal rock dynamic disasters. J China Univ Min Technol 35(03):302–305
28. Jiang YD, Zhao YX, Liu WG et al (2005) Study on 3D model of flat motion instability in deep coal seam roadway. Chin J Rock Mech Eng 24(16):2864–2869
29. Young PR (1993) Rockbursts and seismicity in mines. A. A. Balkema, Rotterdam, pp 23–50
30. Mendecki AJ (1997) Seismic monitoring in mines. Chapman and Hall, London
31. Luo X, Hatherly P (1998) Application of microseismic monitoring to characterise geomechnics conditions in longwall mining. Explor Geophys 13:489–493
32. Hatherly P, Luo X, Dixon R (2001) Seismic monitoring of ground caving processes associated with longwall mining of coal. In: Gibowicz SJ, Lasocki S (eds) Proceedings of the 4th international symposium on rockbursts and seismicity in mines. A. A. Balkema, Rotterdam, pp 121–124
33. Zhang CQ, Feng XT, Zhou H et al (2012) A top pilot tunnel preconditioning method for the prevention of extremely intense rockbursts in deep tunnels excavated by TBMS. Rock Mech Rock Eng 31(3):289–309
34. Abuov MG, Ermekov TM (1989) Studies of the effect of dynamic processes during explosive break-out upon the roof of mining excavations. J Min Sci 24(06):581–590
35. Cai M (2008) Influence of stress path on tunnel excavation response numerical tool selection and modeling strategy. Tunn Undergr Space Technol 23(06):618–628
36. Tang CA, Wang JM, Zhang JJ (2010) Preliminary engineering application of microseismic monitoring technique to rockburst prediction in tunneling of Jinping II Project. J Rock Mech Geotech Eng 02(03):193–208
37. Luo X, Hatherly P, Ross J (2000) Microseismic mapping of floor fracturing for longwall planning at South Blackwater Colliery. Rockburst and Seismicity in Mines-RaSiM5, pp 337–342
38. Jiang FX, Luo X, Yang SH (2003) Microseismic detection study of spatial rupture of overburden and mining stress field. Chin J Geotech Eng 25(01):23–25
39. Jiang FX, Yang SH, Luo X (2003) Microseismic monitoring reveals the spatial rupture morphology of the surrounding rock in the mining field. J Coal 28(04): 357–360
40. Ye GX, Hatheply P, Jiang FX et al (2009) Application of geophysical logging technology in coal mine rock mass engineering exploration. Chin J Rock Mech Eng 28(07): 1342–1352
41. Wang G, Dou L, Li Z et al (2014) Analysis of the spatial breeding mechanism of impact ground pressure and its microseismic characteristics. J Min Saf Eng 31(01):41–48
42. Wu A, Wu L, Liu X et al (2012) Spatiotemporal distribution of mine microseismic activity. J Univ Sci Technol Beijing 34(06):609–613
43. Cheng Y, Jiang F, Cheng J et al (2006) Preliminary study on microseismic detection of mine earthquakes induced by key layer movement. J China Coal Soc 31(03):273–277
44. Cao A, Luo X, Dou L (2011) Experimental study on microseismic effect of shock vibration wave propagation in mining coal and rock mass. J Min Saf Eng 28(04):530–535
45. Jiang F, Zhang X, Yang S et al (2006) Discussion on the spatial structure of overburden in longwall mining field. Chin J Rock Mech Eng 25(05):979–984
46. Sun Z, Zhu D (1993) Drilling hole double-end sealing leak detection device. China: 90225165.03.31
47. Liang S, Li X, Mao Y et al (2013) Time-domain characteristics of overlying strata failure under condition of longwall ascending mining. Int J Min Sci Technol 23(02):207–211
48. Zhang Y, Peng SS (2000) Design considerations for tensioned bolts. In: Proceedings of the 21st international conference on ground control in mining, Morgantown, West Virginia, pp 131–140

References

49. Peng C (2015) Review of the latest progress in research hotspots of mine pressure and strata control. J China Univ Min Technol 44(01):1–8
50. Peng SS (2008) Coal mine ground control, 3rd edn. Syd Peng Publisher, Morgantown
51. Peng SS (2007) Ground control failures. Syd Peng Publisher, Morgantown
52. Anno (1999) Time-dependent behaviour of deep level tabular excavations in hard rock. Rock Mech Rock Eng 32(2):123–155
53. Pan Y, Wang Z, Zhang Y (2008) Application of catastrophe theory in dynamic instability of rock system. Science Press, Beijing
54. Xie H, Ju Y, Li L (2005) Rock strength and overall failure criterion based on energy dissipation and release principle. Chin J Rock Mech Eng 24(17):3003–3010
55. Pan Y, Wang Z (2004) Functional increment of rock mass dynamic instability—a study on the sudden change theory method. J Rock Mech Eng 23(09):1433–1438
56. Obert L, Duvall WI (1941) Use of subaudible noises for prediction of rockbursts II-report of investigation. U.S. Bureau of Mines, Denver
57. Cao W, Zhao M, Tang X (2003) Statistical damage simulation study of rock fracture process. Chin J Geotech Eng 25(02):184–187
58. Ni X, Zhu Z, Zhao J et al (2009) Digitalized microscopic damage mechanics test study of rock fracture process. Rock Soil Mech 30(11):3283–3290
59. Zhang M, Jiang Q, Feng X et al (2018) Triaxial compression test and reinforcement mechanism analysis of fractured marble anchor grouting specimen. Rock Soil Mech 39(10):3651–3660
60. Tang C, Zhao W (1997) RFPA2D, a software system for analysis of the whole process of rock fracture. J Rock Mech Eng 16(05):507–508
61. Tang CA (1997) Numerical simulation of progressive rock failure and associated seismicity. Int J Rock Mech Min Sci 34(02):249–261
62. Huang M, Tang C, Zhu W (2000) Numerical simulation study of rock fracture process. J Rock Mech Eng 19(04):468–471
63. Bo Q (2013) Research on deformation and failure law of deep tunnel surrounding rock and its application based on ABAQUS. Qingdao Technological University
64. Liu B, Liu Q (2011) Study on the law of microseismic activity during the incubation and occurrence of rock burst. J Min Saf Eng 28(02):174–180
65. Zhao X, Li Y, Yuan R et al (2007) Study on the dynamic evolution process of rock crack based on acoustic emission location. J Rock Mech Eng 26(05):944–950
66. Cao S, Liu Y, Li Y et al (2009) Experimental study on acoustic emission characteristics of coal rock under different confining pressures. J Chongqing Univ 32(11):1321–1327
67. Zhao H, Yin G (2011) Experimental study on acoustic emission characteristics and damage equation of gas-bearing coal. Rock Soil Mech 32(03):667–671
68. Zuo J, Xie H, Wu A et al (2011) Study on the failure mechanism and mechanical properties of deep coal rock monomer and combination. J Rock Mech Eng 30(01):84–92
69. Shkuratnik VL, Filimonov YL, Kuchurin SV (2005) Experimental regularities of acoustic emission in coal samples under triaxial compression. J Min Sci 41(1):44–53
70. Shkuratnik VL, Filimonov YL, Kuchurin SV (2006) Experimental acoustic emissive memory effect in coal samples under triaxial axial-symmetric compression. J Min Sci 42(03):203–210
71. Shkuratnik VL, Filimonov YL, Kuchurin SV (2004) Experimental investigations into acoustic emission coal samples under uniaxial loading. J Min Sci 40(05):458–464
72. Voznesenskii AS, Tavostin MN (2005) Acoustic emission of coal in the post limiting deformation state. J Min Sci 41(4):291–298
73. Dou L, He X (2001) Theory and technology of impact mining pressure control. China University of Mining and Technology Press, Xuzhou
74. Qi Q, Lei Y, Li H et al (2007) Theory and practice of preventing and controlling impact ground pressure by deep hole roof blasting. J Rock Mech Eng 26(S1):3522–3527
75. Zhang M (2001) Prediction and control of impact ground pressure in China. J Liaoning Tech Univ 20(4):434–435
76. Wen Z (2011) Research on the mechanical model and key technology of controlling the roadway along the goaf without coal pillar. Shandong University of Science and Technology, Qingdao

77. Wen Z, Jiang Y, Song Z et al (2011) Research on the catastrophe system and control mechanical model of the surrounding rock structure of the roadway along the goaf. J Hunan Univ Sci Technol (Nat Sci Ed) 26(03):12–16
78. Jiang F, Yao S, Wei Q et al (2015) Research on the mechanism and application of mine-induced impact ground pressure field warning. J Rock Mech Eng 34(S1):3372–3380
79. Jiang F, Feng Y, Liu Y (2014) Research on the dynamic assessment method of impact danger before mining. J Rock Mech Eng 33(10):2101–2106
80. Miao X, Sun H, Wu Z (1999) Analysis of the mechanism of impact mining pressure in the soft rock mining area in the eastern part of Xuzhou. J Rock Mech Eng 18(04):428–431
81. Zhou G, Liu W, Jiang Y et al (2008) Analysis of energy accumulation and release characteristics of impact ground pressure in mining area. J Min Saf Eng 25(01):73–77
82. Dou LM, Lu CP, Mu ZL et al (2009) Prevention and forecasting of rockburst hazards in coal mines. Min Sci Technol 19(05):585–591
83. He H, Dou L, Gong S et al (2010) Study on the law of shock induced by the movement of key strata in overburden. Chin J Geotech Eng 32(08):1260–1265
84. Dou L, He H (2012) Study on the evolution law of OX-F-T spatial structure of overburden in coal mine. Chin J Rock Mech Eng 31(03):453–460
85. Xie G (2005) Mechanical characteristics of macro stress shell of fully mechanized mining face and its surrounding rock. J Coal 30(03):309–313
86. Xie G, Yang K (2010) Evolution characteristics of macro stress shell of surrounding rock in mining field. Chin J Rock Mech Eng 29(S1):2676–2680
87. Yu B (2013) Prevention and control technology of mining shock pressure in Jurassic coal seam group of Datong Mining Area. Coal Sci Technol 41(09):62–65

Open Access This chapter is licensed under the terms of the Creative Commons Attribution-NonCommercial-NoDerivatives 4.0 International License (http://creativecommons.org/licenses/by-nc-nd/4.0/), which permits any noncommercial use, sharing, distribution and reproduction in any medium or format, as long as you give appropriate credit to the original author(s) and the source, provide a link to the Creative Commons license and indicate if you modified the licensed material. You do not have permission under this license to share adapted material derived from this chapter or parts of it.

The images or other third party material in this chapter are included in the chapter's Creative Commons license, unless indicated otherwise in a credit line to the material. If material is not included in the chapter's Creative Commons license and your intended use is not permitted by statutory regulation or exceeds the permitted use, you will need to obtain permission directly from the copyright holder.

Chapter 2
Research on Spatio-Temporal Migration Law of Overburden Structure in Mining Area

The coal mine mining area is always in the process of continuous advancement and development, and the mining stress is also in the process of continuous development and change with the advancement of the mining area. In view of the engineering characteristics of the continuous advancement of the coal mine mining area, the spatio-temporal migration of the overburden in the mining area, mining stress, etc. are all in the process of continuous development and change, and this change is regular and determined by the movement of the rock layer.

2.1 Construction of Overburden Spatial Structure in Mining Area

Mining causes a large range of movement and stress redistribution in the overlying rock layer, especially the destruction and change of the coal and rock body around the mining area. The mining of coal seams causes the spatial structure of the overburden in the mining area to break and move, accompanied by the development and change of stress in the surrounding coal and rock body, further inducing dynamic disaster accidents in the mining area. The three-dimensional space of the overburden in the mining area refers to a series of mechanical and mathematical models established to describe the spatial structure of the mining area [1]: Taking the coal seam as the reference plane, the space where the physical form of the floor and roof changes caused by the advancement of the mining area, including the vertical space from the bottom of the coal seam to the ground in terms of longitudinal height, the horizontal space in the inclined direction and advancement direction of the mining area based on the coal seam as the reference plane, and the time space experienced from the start of the working face to the stop line from the perspective of mining stress change transfer.

2.1.1 Overview of the Three-Dimensional Spatial Structure Model of the Mining Area

The three-dimensional spatial structure of the mining area extends the traditional two-dimensional space of the mining area to the three-dimensional space with the advancement of the mining area as the main line. Experts have conducted in-depth research on it, relying on existing research, the spatial structure of the mining area where the mine pressure of the mining area is manifested is divided into subsidence rock layer, fracture rock layer, collapse rock layer, coal seam, and floor rock layer five groups, as shown in Fig. 2.1.

1. Division of the spatial structure of the mining area where the mine pressure of the mining area is manifested

Mining subsidence refers to the rock layer group that undergoes significant deformation, located on the outside of the mining fracture arch affected by coal seam mining. It includes the three-dimensional spatial structure formed from the edge of the mining stress around the coal seam to the edge of the surface subsidence (excluding the rock layer inside the fracture arch formed by the mining field) (as shown in Fig. 2.2). This space includes the bending subsidence zone in traditional concepts (the space from the vertical outside of the fracture arch directly above the mining field to the surface) A, and also includes the space of the B part around the mining field. Surface subsidence is the direct manifestation of mining subsidence rock layers on the surface. It is affected by various factors such as the burial depth of the coal seam, the height of the fracture arch above the mining field, and the thickness of the topsoil layer. It is a direct reflection of coal mining geological disasters on the surface. Surface subsidence has devastating damage to surface buildings and surface crops, and has a significant impact on rivers, lakes, and embankments. Effectively controlling the range and magnitude of surface subsidence is a topic of close concern in today's mining industry. This part of the rock layer only undergoes bending deformation in shape, no cracks appear throughout the thickness of the rock layer, and no fracture

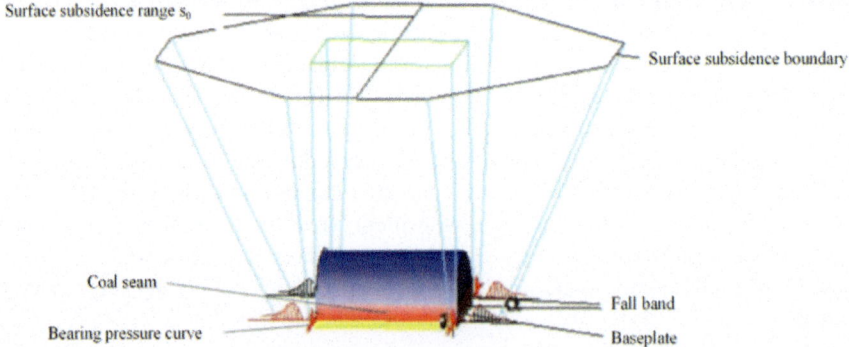

Fig. 2.1 Schematic diagram of the spatial structure of the mining area

2.1 Construction of Overburden Spatial Structure in Mining Area

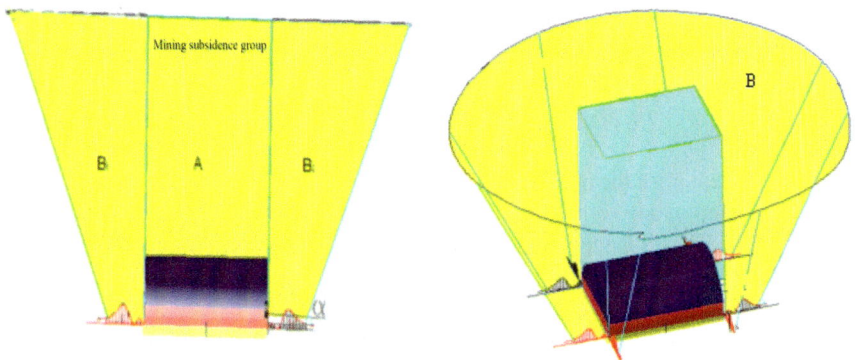

Fig. 2.2 Schematic diagram of the cross-section of the mining subsidence group stratum

occurs. Under the action of gravity, the original cracks and joints in the rock layer may expand, but the expansion of the joint cracks does not penetrate the entire thickness of the rock mass in the vertical direction, and does not reach large-scale lateral communication in the horizontal direction.

The mining subsidence rock layer above the mining field generally has the following judgment characteristics: the rock layer does not have obvious through cracks in appearance, and basically maintains the original state of the rock layer. The rock layer maintains the original mechanical properties of the rock mass. The span of the rock layer in the air (L_{Mi}) is less than the first fracture step of the rock layer $(C_{0(Mi)})$, that is, $L_{Mi} < C_{0(Mi)}$. There may be tiny regenerated cracks at the bottom of this rock layer group, but the cracks cannot conduct water flow and air flow. Studying and analyzing the characteristics of mining subsidence rock layer changes and the range of mining subsidence is of great significance for understanding the development direction of fracture arches and safety accidents such as roof water disasters.

The fractured rock layer refers to the rock layer group that has larger cracks or even fractures directly below the mining subsidence rock layer, directly above the mining field caving zone. It generally has important features of water conduction and gas conduction. The fractured rock layer generally has the following judgment characteristics: after the rock layer sinks, it can be distinguished from the arrangement of the rock blocks, and the rock blocks in the general fracture group have obvious regularity. After the rock layer sinks, the rock blocks can always maintain the transmission of horizontal force in the horizontal direction. The allowable sinking space of the rock layer S_A is less than the thickness of the rock layer itself h_{Mi}, and the span of the rock layer in the air (L_{Mi}) is greater than the first fracture step of the rock layer $(C_{0(Mi)})$, that is, $S_A + S_{A'} < h_{M6}$ and $L_{Mi} > C_{0(Mi)}$. The rock layer has obvious through cracks, and the cracks penetrate the entire thickness of the rock mass in the vertical direction, and have obvious characteristics of conducting water and air flow in the vertical direction. The movement of each transmission rock beam is the main source of pressure on the mining field support and the pressure in the "internal stress field" in front of the coal wall. Studying the fractured rock layer, especially the study of the development law of the fracture arch, is of great significance for roof water penetration and other roof accidents.

The caving rock layer refers to the rock layer group above the mining field coal seam, directly below the fractured rock layer. An important feature of this type of rock layer is that after the rock layer collapses (collapses), it cannot always maintain the transmission of horizontal force in the horizontal direction. The caving rock layer directly affects the advancement and support status of the working face, and also has a greater impact on the roadway support on both sides of the mining field. To judge whether a rock layer above the mining field belongs to the caving group, it has the following judgment characteristics: after the rock layer sinks, it can be distinguished from the arrangement of the rock blocks, and the rock blocks of the general caving group have no obvious regularity. After the rock layer sinks, the rock blocks cannot always maintain the transmission of horizontal force in the horizontal direction. The allowable sinking space of the rock layer S_A is greater than the thickness of the rock layer (or stratification) h_{Mi}. The caving group rock layer has engineering characteristics such as expansion, water conduction, and gas conduction.

The coal seam refers to the coal and gangue layers sandwiched between the top and bottom rock layers. The coal seam is the main part of the coal series, and the number of coal seams, thickness and its changes are the main factors for evaluating the mining value of the coal field. Commonly used are the range of mining stress distribution (S_X) and the position of the peak mining stress (S_1) and other parameters.

Under the influence of mining pressure in the mining area, the floor of the coal seam undergoes a repeated process of compression, expansion, and re-compression. And around the mined-out area, there is always a large shear force. Under the repeated action of these forces, the floor rock mass will move and deform, causing new cracks to appear inside the rock mass and further enlarging the original cracks, and then these cracks connect to form connected cracks, leading to the destruction of the floor.

2. Development law of overburden movement in the mining area

Although we have divided the overburden situation in the mining area based on the coal seam of the working face and the situation of the roof and floor after mining, due to the different heights of the overlying rock layers from the mining area, their lithology and thickness are also different, and their full range of movement and impact on the mining area are also different. Scholars at home and abroad have made great contributions to the study of the structure and migration law of the overlying rock layer, providing theoretical basis and technical support for the control of coal mine disaster accidents.

Practice has proved that in the mining area covered by general rock layers, when the mining depth exceeds a certain value and the width of the working face reaches 150–200 m or more, when the mining area advances beyond the width of the working face, the range of overburden involved in movement and damage due to mining influence and the redistributed stress field are shown in Fig. 2.3. That is, the structural mechanical model of the mining area formed under this mining condition is composed of two parts: the range of the overlying rock layer affected by mining and the range of

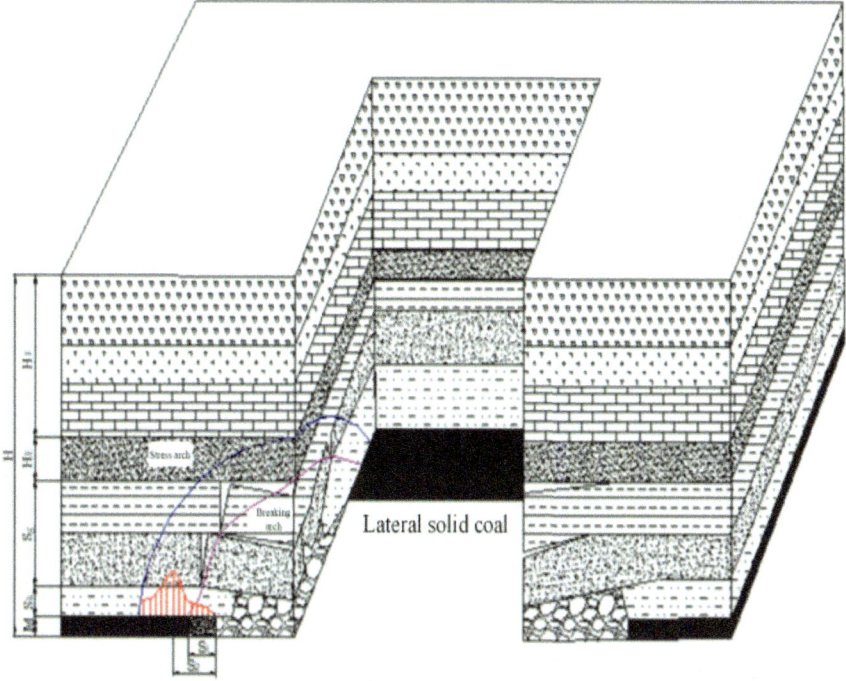

Fig. 2.3 Schematic diagram of the mechanical model of the mining area

the redistributed stress field acting on the coal seam. In order to study the development law of overburden movement in the mining area, control the rock layer movement that has a significant impact on the mining area, and further start from the longitudinal and advancing direction, the morphological structure and mechanical characteristics of each component are described separately [2].

A) Longitudinal development law of overburden in the mining area

The longitudinal movement of the rock layer generally first follows the advancement of the mining area. Under the action of gravity, the rock layer is exposed to a certain span and bends and settles to a certain value. The weak interlayer or contact surface with low strength is destroyed under the action of axial shear stress, causing delamination, and creating conditions for the free settlement and upward movement of the lower rock layer; then after delamination, it is recombined into a pseudo-plastic transfer rock beam that moves simultaneously or almost simultaneously during movement, and finally the settlement value exceeds the allowable limit and collapse occurs. Each rock layer is affected by the advancement of the mining area, and its exposure time, exposure span, and external load from bottom to top are different. Generally speaking, the rock layer at the bottom is exposed first, and the closer to the top, the later the exposure; the exposure span of each rock layer decreases from bottom to top. Because the exposure span of the rock beam has a rule of decreasing from bottom to top, and the size of the shear stress is proportional to the exposure span of the rock beam, the size of the shear stress also decreases from bottom to top. In addition, another important reason for the formation of rock layer longitudinal movement from bottom to top is: the external load acting on each rock layer decreases from bottom to top. In general, The overall trend of rock layer longitudinal movement is to develop from bottom to top, and the combination of upper and lower rock layer movement after delamination is determined by the difference in lithology, thickness, and fracture development of the rock layer; the thickness of the rock layer is more important than the lithology in influencing the delamination and movement combination of the rock layer.

B) Development law of overburden in the advancing direction of the mining area

With the advancement of the mining area, the mining stress and the pressure on the support in front of the coal wall are constantly changing. The development and change law of the mining pressure manifestation in the mining area is determined by the movement development law of the overlying rock layers that affect it. In addition to the influence of the longitudinal development law of rock layer movement, it is more importantly affected by the development law in the advancing direction. During the advancement of the mining area, due to the different mining pressures borne by the overlying rock layers and the differences in support (constraint) conditions, the development of the overlying rock layer movement in the mining area can be divided into the first movement and periodic movement stages:

From the beginning of the rock layer exposure from the cutting eye, to the end of the first fracture movement of one or two transfer rock beams that have a significant impact on the working face, it is the first movement stage (as shown in Fig. 2.4a, b). This includes the first direct fall across the rock layer. The ends of the rock layer at this stage are supported by the coal wall, and its stress state can be regarded as a fixed beam. The first movement of each rock layer in the mining area is called the initial pressure of the mining area. Because the step distance of any rock layer's first movement is much larger than the step distance under normal conditions, the area of the first movement pressure is large, the strength is high, and it may be accompanied by dynamic pressure shock.

From the end of the first movement of each layer to the completion of the working face, the roof rock layer moves regularly and breaks according to a certain period, which is called the periodic movement stage (as shown in Fig. 2.4 c $\sim$ f). In this development stage, the constraint conditions of the rock layer have fundamentally changed: the direct top rock layer in the mining area is a "cantilever beam" with one end fixed. The rock beams above the direct top are unevenly high transmission rock beams, one end is supported by the coal wall, and the other end is supported by the goaf gangue. At this time, the movement step distance is much smaller than the first movement step distance.

In the above two development stages, the movement of the rock layer will go through two development processes: relative stability and significant movement. We call the process where the movement of the rock beam is small and the impact on the mining pressure of the mining area is not obvious, as the rock beam is in a

2.1 Construction of Overburden Spatial Structure in Mining Area

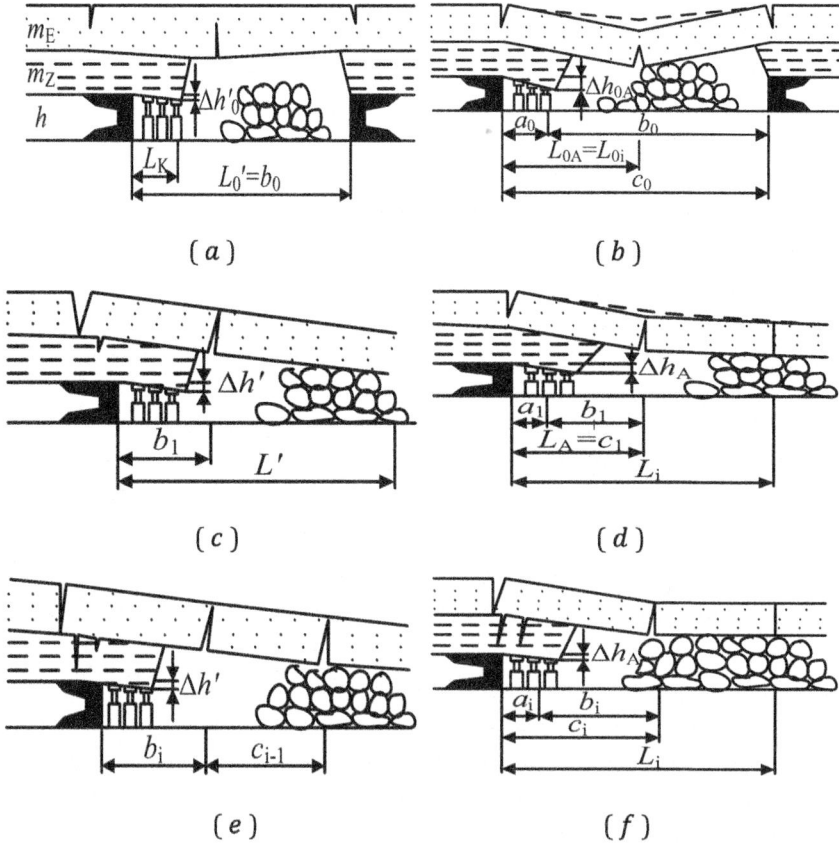

Fig. 2.4 Development process of rock layer movement in the advancing direction

relatively stable process. The parameter describing the length of this process is the relative stability step distance of the rock beam, that is, the distance the working face advances when the rock beam is in a relatively stable state, represented by b, as shown in Fig. 2.4 a, c, e.

The process where the movement of the rock beam is large and the impact on the mining pressure of the mining area is very obvious is called the rock beam is in a significant movement process, that is, the so-called "pressure" process. The parameter describing this movement process is the significant movement step distance of the rock beam. That is, from the beginning of the large-scale movement of the rock beam to the end of the movement, the distance the working face advances, represented by a, as shown in Fig. 2.4 b, d, f.

Where: m_E is the basic roof; m_Z is the immediate roof; h is the coal seam; L_K is the control roof distance; L_0' is the maximum span of the rock beam; $\Delta h_0'$ is the minimum subsidence of the mining area roof; Δh_{OA} is the roof subsidence at the end of the pressure; a_0 is the significant movement step distance of the rock beam; b_0 is the relative stability step distance of the rock beam; c_0 is the initial pressure step distance of the rock beam; L_{OA} is the span of the rock beam at the end of the given deformation pressure; a_1 is the significant movement step distance of the rock beam; b_1 is the relative stability step distance of the rock beam; c_1 is the initial periodic pressure step distance; L' is the limit span of the rock beam on the eve of the periodic pressure; L_A is the minimum span of the rock beam span at the end of the pressure under the given deformation conditions; Δh_A is the subsidence of the mining area roof under the given deformation working conditions of the support; $\Delta h'$ is the subsidence of the mining area roof on the eve of the periodic pressure; a_i is the significant movement step distance of the rock beam; $|c_i|$ is the periodic pressure step distance; L_i The relatively stable span at the end of the rock beam cycle.

As the working face advances, the mining stress changes continuously, and the distribution of mining stress will appear higher and lower than the original rock stress. Corresponding to different mining depths and coal seam strengths, the mining stress field (support pressure field) that is redistributed in front of the coal wall due to mining influence includes the following three parts:

(1) The "internal stress field", in which the coal seam in this stress field has been damaged under the action of mining stress and has entered a plastic or pseudo-plastic damage state. The magnitude and time of its force are directly affected by the fracture movement of the rock beam in the "fracture arch", as shown in the "S_1" interval in Fig. 2.3. The stress value in this interval is generally lower than the original stress;

(2) The "plastic damage zone", the range in front of the coal wall that enters the plastic damage state under the influence of mining stress, as shown in "S_2" in Fig. 2.3. The mining stress in this interval mainly comes from the action of the rock beam in the "stress arch" of the mining area;

(3) The "elastic compression zone", where the coal seam is in an elastic compression state under the action of mining stress. The stress in this interval is higher than the original stress.

The three types of mining stress mentioned above each have their own conditions of existence. Different coal seams may have different distribution forms under the same mining conditions. Even if the coal seam conditions and mining technology are the same, but the mining depth is different, the working face advances to different parts, and the distribution form of mining stress is often different. It is necessary to clarify the reasons and conditions for the various forms of mining stress, which is of great significance for mine pressure control, especially for solving the problems of roadway mine pressure control.

2.1 Construction of Overburden Spatial Structure in Mining Area

3. Structural characteristics and evolution laws of "fracture arch" and "stress arch"

In order to effectively control coal mine disaster accidents, scholars at home and abroad have successively proposed many classic models of mining field overburden spatial structure hypothesis, and each has explained certain phenomena, but the construction of models is mainly static models. Based on the research of existing scholars, the author has systematically analyzed the spatial structure of the mining field overburden and the evolution law of mining stress, and constructed a dynamic spatial structure model (Fig. 2.3) that describes the mining field overburden space structure with "fracture arch", "stress arch" and "internal and external stress field" as the core. This structure model describes the overburden movement damage and mining stress magnitude, distribution and its development law with the advancement of the mining field under different mining conditions (including mining height, working face length and mining procedures, etc.) for coal seams under predetermined conditions of different mining depths and rock layer structures. The rock layer above the coal seam can be divided into two parts: the overburden space structure and the outside of the overburden space structure. The outside of the overburden space structure refers to the rock layer that has not undergone obvious movement outside the "fracture arch". The overburden space structure is composed of moving rock layer structures that have a direct impact on the mine pressure of the mining field. As the working face advances, the exposed space of the mining field continues to increase, the overlying rock layer continues to fracture and collapse, and the fracture position is successively misaligned from bottom to top, forming a "fracture arch". At the same time, the stress in the spatial structure surrounding rock is redistributed, and the gravity of the overlying rock layer originally borne by the mined coal body of the working face is loaded on the coal (rock) body on both sides. If the total stress borne by the coal (rock) body exceeds its strength, it will be damaged, and the stress peak will shift to the inside. Each rock beam fracture is accompanied by this process, forming a "stress arch" composed of mining stress peaks of each rock layer outside the "fracture arch", which continuously develops upward in a parabolic shape in the vertical plane of the mining direction and inclination. From the perspective of stress field distribution and structural development, the composition of the mining field spatial structure model is analyzed. Perpendicular to the direction of the working face advancement: ① Longitudinally form a "stress arch" and "fracture arch"; ② Laterally form an "internal and external stress field" and a transfer rock beam.

A) Evolution process of "fracture arch" and "stress arch"

As the working face continues to advance, the range of spatial overburden structure is constantly expanding. In the process of spatial overburden structure continuously developing towards the advancing direction and longitudinal space, the mining field spatial structure develops into two structural mechanical forms, that is, a stress shell (sectional manifestation as "stress arch") is formed on the periphery of the spatial overburden structure, and the "fracture arch" containing the spatial overburden structure. The "fracture arch" is composed of fractured rock beams inside the "fracture arch" that have a significant impact on the manifestation of mine pressure in the mining field. The structure inside the "stress arch" is composed of rock layers within the range of the "external stress field" layer by layer. According to the development process of the mining field overburden structure, the development and change of mining stress can be divided into two major stages, which can be specifically divided into four small stages.

The first stage: the development stage of "fracture arch" and "stress arch", that is, the arching stage.

The working face starts to advance from the position of the cutting eye. When the overlying rock layer of the mining void area has not yet collapsed or the collapsed rock layer has not yet filled the mining void area, that is, the thickness of the collapsed rock layer in the mining void area $h < (m - S_{sink})/(k - 1)$ At this time, the movement of the overlying rock in the mining field can be regarded as hanging in the center of the mined-out area, and the bending structure of the coal rock mass around it is acting. If the bending value is greater than its deflection limit, the rock layers will fracture from bottom to top in sequence. The coal rock mass on both sides of the fracture not only bears the weight of the overlying rock layer, but also carries the overburden load borne by the original excavation area, which is very likely to cause compressive damage, causing the elastic stress peak to move outwards, forming an "external stress field".

The development and change of mining stress at this stage can be further divided into three stages:

Stage I Coal wall elastic stage—always maintain the ability to support the overlying rock layer;

In the initial mining stage, starting from the cutting eye, as the mining field continues to advance, the exposed space of the mined-out area continues to increase, and the overburden load borne by the coal body in the original mining area is transferred to the coal walls on both sides through the rock beam. The load borne by the coal wall will gradually increase. At this time, the coal wall is in the elastic stage, and due to stress concentration, the peak value of mining stress is at the edge of the coal wall.

Stage II Coal wall damage and fracture destruction—the peak of elastic stress is transferred outward, and the support capacity of the coal wall is greatly reduced;

After the coal seam is excavated, the original three-dimensional mechanical equilibrium state is broken, and its mechanical parameters will be weakened after unloading, so its support capacity will be reduced. The coal body near the coal wall is prone to damage and fracture due to the continuous advancement of the mining field. The overburden load borne by the coal body in the original mining area is likely to exceed the load limit borne by the coal wall, causing damage and fracture. At this time, in order to continue to maintain the stage of stable equilibrium of the overlying rock layer in the mining field during the advancement process, the coal body outside the coal wall that has not been damaged or has less damage will bear the overburden load, causing the peak of elastic stress to shift outward.

Stage III "Internal and external stress field" formation stage;

When the exposed space of the mining area reaches a certain range, that is, when the exposed length in the advancement direction reaches the first rock beam fracture position of the overburden, the first pressure comes. At this time, the mining stress distribution is clearly divided into two parts with the fracture line as the boundary, that is, the "internal stress field" determined by the "fracture arch" internal fracture rock beam between the fracture line and the coal wall, and the "external stress field" determined by the "stress arch" internal overburden load outside the fracture line. As the rock beams in the "fracture arch" fracture one after another, the overburden load borne by the "internal stress field" gradually increases. This process continues until the "fracture arch" is formed; the "external stress field" determined by the "stress arch" decreases due to the fracture of the unfractured rock beam in the "fracture arch" acting on it, and the load decreases, so the peak of elastic stress decreases in stages, showing a "internal ("internal stress field") large (range, peak) external ("external stress field") small (peak)" rule, manifested as fluctuating changes.

The second stage: the formation stage of the "fracture arch" and "stress arch", that is, the arching stage.

As the working face continues to advance, the collapse range of the overlying rock layer in the mining field continues to develop in the direction of the working face and the longitudinal direction of the space. When the thickness of the collapsed rock layer in the mined-out area is $h \geq (m - S_{sink})/(k - 1)$, the fallen rock fragments have filled the mined-out area, and the overlying rock layer in the mining field no longer fractures due to the lack of collapse space, and the movement state presents a slow sinking towards the mined-out area. At this time, the movement state of the overlying rock layer in the mined-out area of the mining field can be regarded as a plate bending structure located on different foundations around and in the center, that is, located on the structure of the fallen gangue and the surrounding coal rock mass in the mined-out area. In the formula, h is the thickness of the collapsed rock layer, m is the mining thickness, k is the collapse expansion coefficient of the mined-out area, S_{sink} is the sinking size of the overlying rock beam. After that, as the working face continues to advance, the overlying unfractured rock layer in the mined-out area moves and destroys in the "plate-shell" structure in space, and gradually develops in the direction of the working face and the longitudinal direction of the space, until the collapse expansion coefficient of the mined-out area reaches the minimum. At this time, the overlying rock layer has no space to continue sinking, reaching full mining. This stage is the formation stage of the "fracture arch" and "stress arch".

At this time, the development and change of mining stress enters the fourth stage, Stage IV "Internal stress field" disappearance stage.

"After the "fracture arch" is formed, the "internal and external stress fields" reach an ideal stable state. The upper "support layer" bends under the load it carries, part of the weight acts on the fractured rock beam inside the "fracture arch", causing the "internal stress field" to continue to increase under load, the coal body within the range is severely damaged (the experiment manifests as continuous extrusion of foam debris), and the bearing capacity is significantly reduced; the coal body within the "external stress field" is reduced under load, because the coal body within the range is less disturbed, the damage is also smaller, the bearing capacity is almost unchanged, and the peak value of elastic stress continues to decrease significantly.

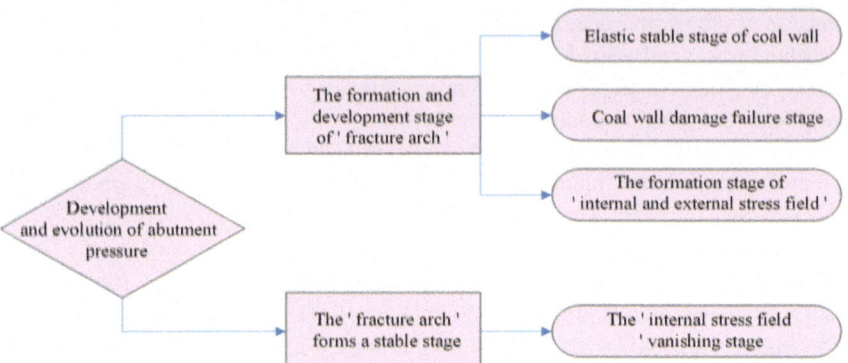

Fig. 2.5 Evolution law of mining stress development

Comprehensive analysis shows that the development and evolution process of mining stress is closely connected with the formation and development process of the dynamic structural mechanics model of the mining field centered on "rock layer movement". The generation, development, and stabilization process of the structural mechanics model is closely related to the evolution process of mining stress, as shown in Fig. 2.5.

B) Mechanical structure characteristics and evolution law of "fracture arch"

The mining process is divided into two stages: ① the insufficient mining stage, that is, the working face advancement distance L_X < working face width L_0; ② the sufficient mining stage, that is, the working face advancement distance L_X > working face width L_0.

In the insufficient mining stage, the height of the overburden spatial structure in the mining field generally develops linearly with the advancement of the working face, continuously developing forward in the direction and upward in space, and the height of the spatial structure is about half of the short span of the mined-out area [3]. However, this development law is conditional. The previous analysis shows that the overburden spatial structure of the mining field is mainly determined by the width of the working face. When the width of the working face is fixed, the maximum development height of the overburden spatial structure is fixed. When the advancement distance of the working face has not reached the width of the working face, the development height of the spatial overburden structure is related to the advancement length of the working face. When the advancement distance of the working face reaches the width of the working face, the development height of the spatial overburden structure is about half of the width of the working face, that is, before the "square" appears in the mined-out area, the development height of the spatial overburden structure increases with the advancement of the working face. After the "square" appears in the mined-out area, the development height of the spatial overburden structure develops to the maximum height under the condition of the working face width.

2.1 Construction of Overburden Spatial Structure in Mining Area

(1) The "fracture arch" is composed of fractured rock beams, and the overburden within the range is the main force source of mining pressure manifestation in the mining field;

(2) The "fracture arch" in the mining field overburden looks like a semi-ellipsoid, the arch base is located at the first rock beam fracture position on both sides of the working face, the arch top is located in the hard rock layer, that is, the "support layer", the height is about half of the working face width;

(3) During the advancement of the mining field, the "fracture arch" continues to develop forward in the advancement direction, and basically remains stable after developing to the limit height ($S_g = L_0/2$) in space.

C) Mechanical characteristics and evolution law of "stress arch"

The rock layer inside the "stress arch" bears and transfers the load of the overlying rock layer, and is the main bearing body. The "fracture arch" structure is located in the pressure relief zone inside the "stress arch". When the overburden structure inside the "stress arch" is unbalanced, major disasters such as impact ground pressure will occur. Therefore, it is necessary to understand the dynamic evolution process of the "stress arch".

Assuming that there are k layers of overburden structure in the mining field, the $n + 1$ layer is the support layer, after the i layer rock beam inside the "fracture arch" is fractured, the original overburden load acting on it is transferred to the outside of the "fracture arch". As shown in Fig. 2.6, the load borne by the unit length of the i layer rock beam outside the "fracture arch" is

$$q_i = q_{1(k-i)} + q_{2(k-i)} = \gamma \cdot H_i \cdot L_i + \gamma \cdot H_i = \gamma \cdot H_i(1 + L_i) \qquad (2.1)$$

where, $q_{1(k-i)}$ and $q_{2(k-i)}$ are the loads borne by the i layer rock beam inside and outside the "fracture arch" respectively; H_i is the burial depth of the i layer rock beam; L_i is the fracture length of the i layer rock beam.

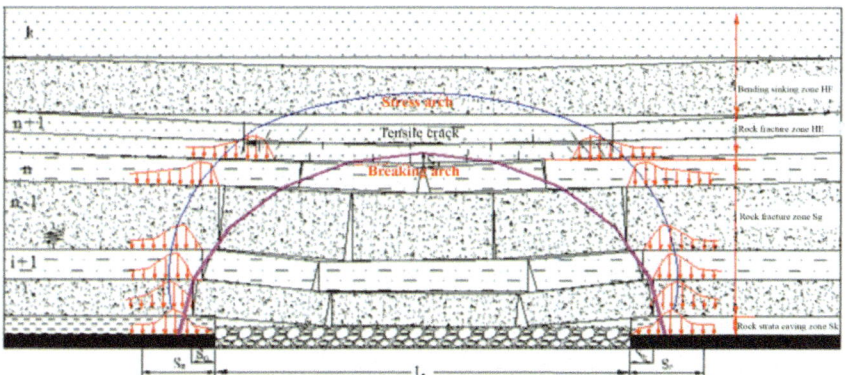

Fig. 2.6 Stress calculation model of the extrados rock beam

When the rock beam undergoes tensile failure, a large number of tensile cracks will occur at the edge of the fracture location, and the load it bears will increase instantaneously, which can easily cause the strength of the rock beam at the fracture location to decrease. The mining stress peak shifts to the outside (see Fig. 2.7a). This situation is likely to occur in deep mines. If the strength of the rock beam is sufficient to support the load transmitted from the "fracture arch", and no failure occurs, then the mining stress peak is at the fracture location (see Fig. 2.7b). This situation is likely to occur in shallow mines.

The width of the "stress arch" is $L_{stress} = L_0 + 2S_e$; the spatial development height is $H_{stress} = L_0/2 + H_{n+1}$.

The distribution state of the "stress arch" is closely related to the overburden rock properties and structure. The ability of the rock layer to resist damage and bear the overburden load is directly proportional to the rock strength. To better understand the "stress arch" and closely integrate it with the field, we divide the overburden structure into four types [2]: "Hard-Hard Type" (JYJY), "Hard-Weak Type" (JYRR), "Weak-Hard Type" (RRJY), "Weak-Weak Type" (RRRR) (Figs. 2.8, 2.9, 2.10 and 2.11).

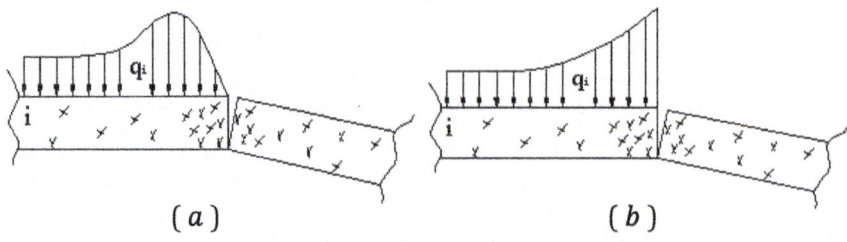

Fig. 2.7 Distribution of mining stress in rock beams

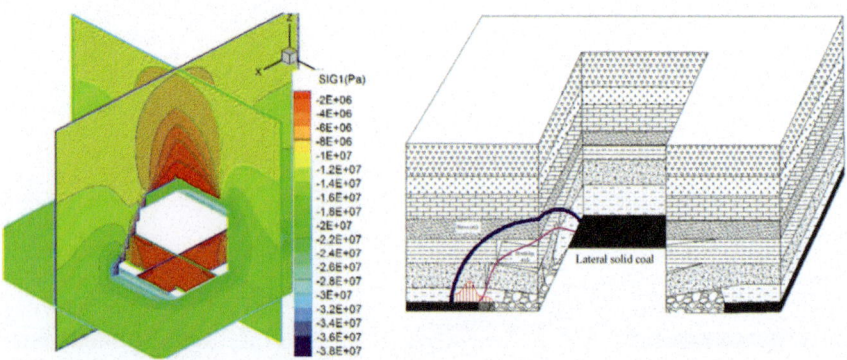

Fig. 2.8 Distribution pattern of JYJY type "stress arch"

2.1 Construction of Overburden Spatial Structure in Mining Area

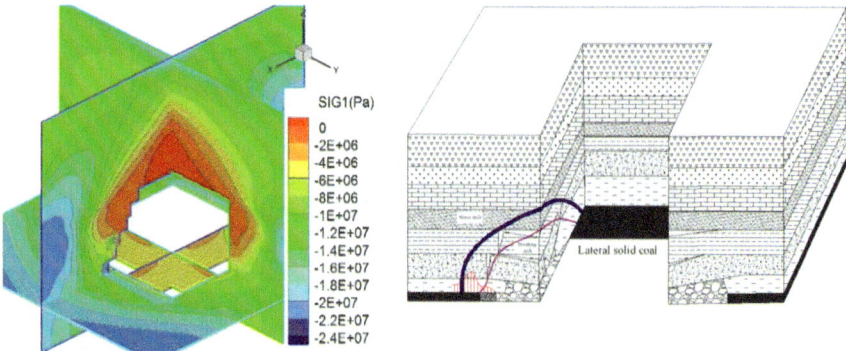

Fig. 2.9 Distribution pattern of JYRR type "stress arch"

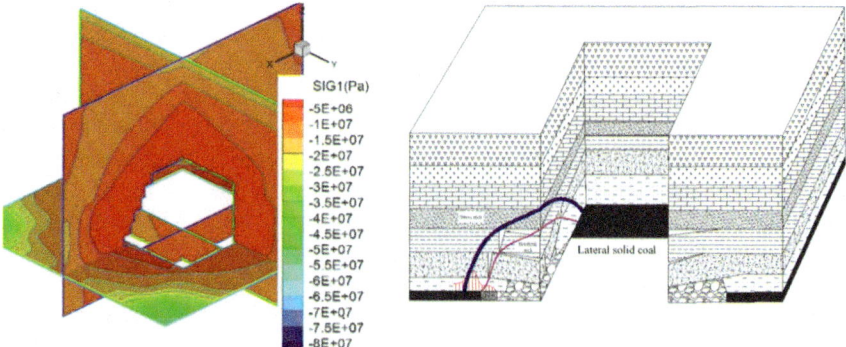

Fig. 2.10 Distribution pattern of RRRR type "stress arch"

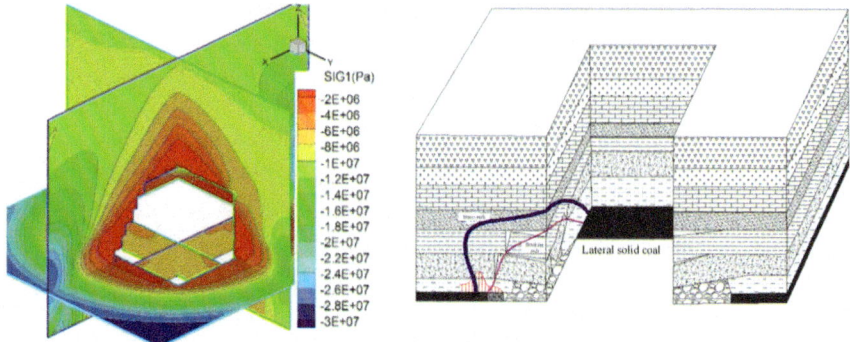

Fig. 2.11 Distribution pattern of RRJY type "stress arch"

The "stress arch" is a set of annular stress envelopes reflecting the stress transmission relationship between rock layers. Its position determines the range of the "external stress field", revealing the range of the force transmitted from the overburden rock layer outside the mining field "fracture arch" to the surrounding rock of the working face. The overburden rock properties of the mining field determine the shape of the "stress arch". The upper limit points of the "Hard-Hard Type", "Weak-Weak Type", and "Weak-Hard Type" are biased towards the mining field, showing a " ⌒ " shape; the "Hard-Weak Type" "stress arch" is in the shape of " ⌂ ". The rock layers that have an impact on the mine pressure manifestation in the mining field are located within the "stress arch".

The "fracture arch" reflects the evolution of the spatial structure movement of the overburden formed by mining. It is composed of fractured rock beams within the "fracture arch" that have a significant effect on the mine pressure manifestation of the mining field. The boundary line is connected by the fracture lines of each rock beam. The rock layers that have a significant impact on the mine pressure manifestation in the mining field are located within the "fracture arch".

The overburden rock properties of the mining field determine the development form of the "stress arch" in the overburden structure, but there are always "four spaces" in the direction of the mining field advancement, namely the goaf compaction zone, the unloading shell, the stress shell, and the original stress zone, as shown in Fig. 2.12a [4]. According to the deformation conditions of the coal body, the coal body can be divided into four areas from the coal wall to the deep part, namely the loosening fracture zone, the plastic strengthening zone, the elastic deformation zone, and the original rock state zone. The "loosening fracture zone" is the commonly referred to "internal stress field". The coal body in the area has been cut into blocky by cracks, the more serious the closer to the coal wall. Its cohesion and internal friction angle are reduced, the strength of the coal body is significantly weakened, and the stress of the coal body in the area is lower than the original rock stress, so it is also called the "unloading zone". The coal body in the "plastic strengthening zone" is in a plastic state, but has a higher bearing capacity. The coal body in the "elastic deformation zone" is still in an elastic deformation state under the action of mining stress, and the stress is higher than the original rock stress. The coal body in the "original rock stress zone" has basically not been affected by mining, and the coal body is in the original rock state. As shown in Fig. 2.12b.

2.1.2 Classification of Overburden Spatial Structure in Mining Area

There are two meanings in the concept of the spatial structure of the overburden in the mining area: one refers to the shape characteristics of the fracture edge of the surrounding rock mass in the mining area; the other refers to the movement structure formed by the rock layers inside the fracture zone. The former (fracture) is the basis

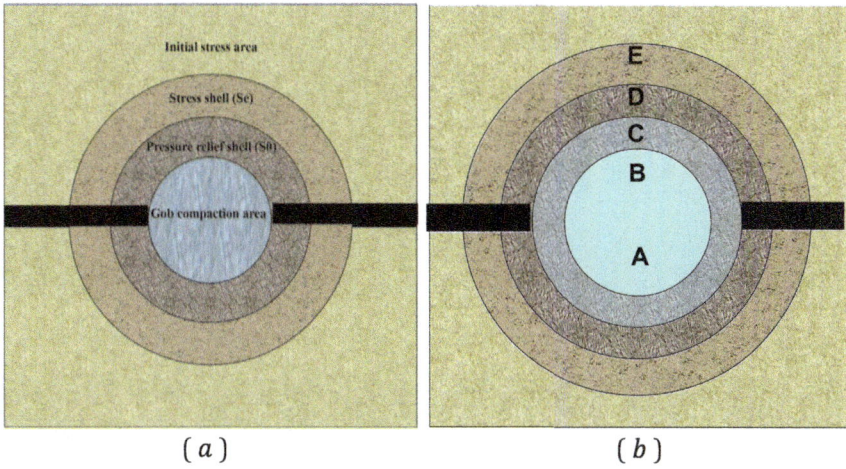

Fig. 2.12 Mechanical model of the "four-layer space" structure of the mining area A-Mining void compaction zone; B-Loose fracture zone; C-Plastic strengthening zone; D-Elastic deformation zone; E-Original rock state zone

for the formation of the latter (structure). The form of the spatial structure of the overburden within the mining area and the mine range is initially determined by the design stage and finally realized by mining activities, which changes with the mining stage. The existing research on the spatial structure of the overburden in the mining area that is widely accepted by experts and scholars is that Professor Jiang Fuxing summarized the three-dimensional structure formed around the mining area after the fracture of different mining boundaries, basic roofs and upper rock layers into the following four types: "θ" type, "O" type, "S" type and "C" type [5, 6].

1. "θ" type spatial structure with "support in the middle"—isolated working face with mining on all sides

The problem of ground pressure control of the "isolated working face with mining on all sides" formed by skip mining and thick coal seam changing from layering to top coal caving (Fig. 2.13a) has become a prominent problem faced by many mines. As can be seen from the figure, in addition to the direct roof and basic roof forces on the coal pillar, the main force is the force of the "overburden spatial structure with support in the middle" formed by multiple layers of overburden. This multi-layer "overburden spatial structure with support in the middle" looks like the letter "θ" in plan projection, and its movement determines the mode and degree of coal pillar damage.

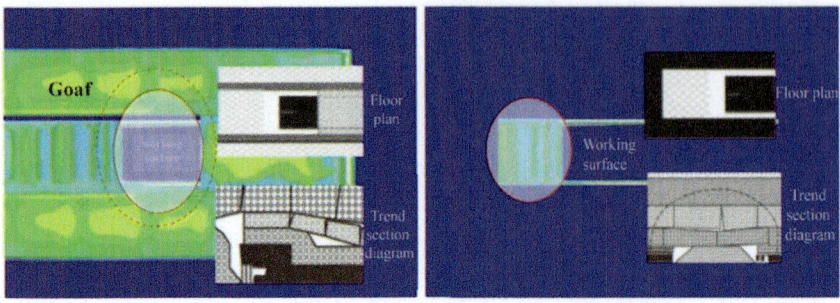

(a) "θ" type overburden spatial structure   (b) "O" type spatial structure

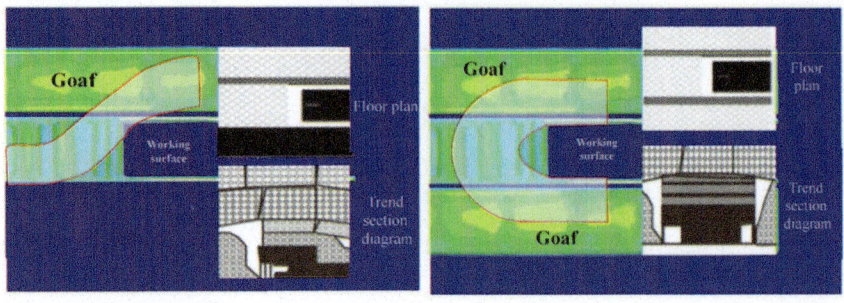

(c) "S" type spatial structure   (d) "C" type overburden spatial structure

Fig. 2.13 Mechanical model of the "four-layer space" structure of the mining area

2. "O" type spatial structure with "no support in the middle"—working face with mining on one side

This type of mining area is generally the first mining face, as shown in Fig. 2.13b, surrounded by solid coal, the range of the spatial structure is closely related to the oblique length of the working face, the composition of the rock layer, the advancement distance, etc. According to the observation results of multiple working faces and simulation experiments, when the working face advances to a certain distance, the maximum fracture height no longer extends, and the scale parameters of the spatial structure enter a relatively stable stage.

3. "S" type spatial structure—working face with mining on two sides

The working face with mining on two sides refers to: one side is the mined-out area after the upper section is mined (only a small coal pillar of 3–5 m is left between it and this working face to protect the roadway), and the other side is the mined-out area of this working face (Fig. 2.13c). After the basic roof of this working face is initially fractured, the upper rock layer will move together with the same layer of rock in the upper section working face, that is, the upper rock layer of the basic roof will move together, forming an "S" type spatial structure. The movement of this spatial structure is the main reason for the difficulty of advanced support in the mined-out side roadway.

4. "C" type spatial structure—isolated working face with mining on three sides

This type of mining area is the isolated working face left by skip mining. In recent years, the flat roadway support of the top coal caving isolated working face in comprehensive mining has become a very difficult task. The advanced mining stress influence distance of some working faces has reached 120 m, and the roadway repair project has become the bottleneck of the working face advancement. There are also large-scale slab problems in deep working faces. Figure 2.13d The morphology of the spatial structure of the overlying strata in this type of mining field is presented. The strata above the three basic goaf areas have merged into one, forming a nearly "C"-shaped spatial structure. The large-scale movement of the "C"-shaped spatial structure is the main reason why the stress influence distance of the isolated working face is 2 to 3 times greater than that of the ordinary working face.

2.2 The Evolution and Development Law of Overburden and Mining Stress in the Mining Field

Mining dynamics is the science of studying the "re"-deformation and "re"-destruction of coal and rock bodies that have undergone deformation and damage during the mining process. Mining stress refers to the force that promotes the surrounding rock to move towards the mined space after mining. This force includes the stress in the surrounding rock related to the mined space and the force of the moving rock layer induced by mining on the boundary of the surrounding rock (the wall of the mined space).

2.2.1 The Evolution and Development Process of Overburden and Mining Stress in the Mining Field

The mining stress in the rock mass around the mining field is the result of the movement of the overburden in the mining field. Its size is related to the mining depth and the exposed area of the rock layer after mining, and its distribution and manifestation are constantly changing [7–10]. The manifestation of mining stress is not entirely dependent on the size of the pressure, but is closely connected with the bearing capacity of the bearing body. Theoretical research and field practice have proven that the changes in mining stress and its manifestation from the start of mining field advancement to the end of the first pressure of the control rock layer (caving rock layer and basic roof) can be divided into three stages. As shown in Fig. 2.14.

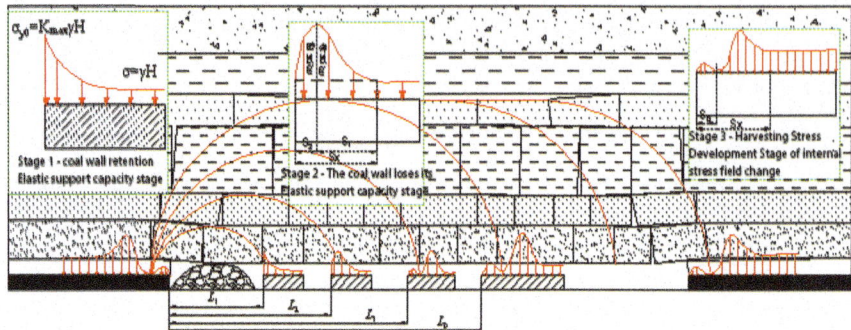

Fig. 2.14 Evolution process of mining stress and mining field structure

1. The first stage—the coal wall maintains its elastic support capacity stage

The mining field starts to advance from the cutting eye, and as the working face advances, the exposed space of the roof also increases, and the roof rock layer will undergo periodic fracture and movement. The pressure transmitted by the roof rock layer to the coal walls on both sides of the mining field also gradually increases. Since the coal wall has a certain hardness and strength, the mining field advances within a certain range, and the pressure transmitted by the roof has not reached the limit of coal body destruction. As shown in the advancement degree L_1 range in Fig. 2.14, the entire coal wall is in an elastic compression state, and the distribution of mining stress is a monotonically decreasing curve with a peak at the coal wall.

In this stage, the distribution range of mining stress S_x is relatively small. The coal wall maintains its inherent support capacity, and the coal wall in front of the mining field is always in an elastic deformation state, and it is not easy for the coal wall to have roof leakage and slabbing phenomena. The mining stress concentrated at the coal wall during this stage continues to increase with the advancement of the mining field, but it has not reached the limit of plastic destruction of the coal body.

2. The second stage—the coal wall loses its elastic support capacity

As the mining field continues to advance, the exposed space of the roof of the working face gradually increases, and the pressure transmitted to the coal wall through the roof also gradually increases. With the increase of the tangential stress of the coal wall, the coal wall reaches its elastic support limit and begins to undergo plastic deformation and even destructive deformation. As the support capacity of the coal wall decreases, the peak of mining stress will gradually shift to the inside of the coal wall until a new stress balance is reached, as shown in the advancement degree L_2 range in Fig. 2.14. This stage starts from the change in the support capacity of the coal wall and ends before the fracture of the lower part of the fracture group. The distribution of mining stress on the coal seam will be divided into 2 intervals: the pressure gradually rises in the plastic zone (the coal body has been completely destroyed), and the pressure monotonically decreases in the elastic zone, and the junction of the elastic–plastic

2.2 The Evolution and Development Law of Overburden and Mining Stress ...

zone is the position of the pressure peak. The distribution range of mining stress S_x is also composed of two parts, the plastic zone S_1 and the elastic zone S_2.

At this stage, under specific roof conditions, it is possible to form a destructive deformation zone—the internal stress field.

From the fracture of the lower part of the fracture group to the touching of the middle part of the rock beam. On the eve of the fracture of the end of the rock beam, the pressure is highly concentrated near the fracture line; the distribution of mining stress in front of the working face is divided along the fracture line of the top rock beam, in terms of size, it is clearly divided into two parts, one is the shear stress reduction area between the coal wall and the fracture line of the fractured rock beam, that is, the "internal stress field" ($\sigma < \gamma H$) determined by the self-weight of the already fractured rock beam between the fracture line and the coal wall, as shown in the vicinity of the L3 range in Fig. 2.14. The "external stress field" ($\sigma > \gamma H$) determined by the overall weight of the overlying rock layer from outside the fracture line of the rock beam to the boundary of the mining stress, the distribution of the mining stress is also divided into two areas, one is the monotonic increase area of the mining stress, that is, the commonly referred to plastic deformation area, the other part is the pressure monotonic decrease area, that is, the elastic deformation area. The formation of the internal stress field originates from the fracture of the roof rock layer, so it is formed simultaneously in front of and behind the mining area, but due to the continuous advancement of the mining area, the changes in the internal stress field in front of the mining area and the changes in the internal stress field behind the mining area have completely different rules.

A) Formation of the internal stress field.

As shown in Fig. 2.15, before the rock beam A fractures, the distribution curve of the mining stress in front of the working face is shown as 1, the peak of the mining stress is concentrated near the fracture line B point, and the clamping force point of the rock beam A is also at the B point. At the moment when the rock beam A fractures, the mining stress undergoes a rapid transformation, with the B point as the boundary, the mining stress is divided into two peaks that move in opposite directions, the front moves to the C point, and the back moves to the D point, as mentioned above, the C point is the boundary point of the elastic–plastic area of the mining stress outside (i.e., the external stress field), and the D point is the pressure peak of the mining stress inside (i.e., the internal stress field). As the rock beam A moves from the initial fracture to the stop, the distribution of the mining stress also slowly changes from curve 2 to the state of curve 3, and the peak will also go through the process of B, C, E changes. At this time, similar changes will occur in the coal wall in front of the working face and the coal wall behind the cutting eye.

B) The development and change rule of mining stress in the coal wall in front of the working face.

After the formation of the internal stress field, a curve as shown in 1 in Fig. 2.15 will be formed in front of the working face, but due to the continuous advancement of the working face, the internal stress field will continue to shrink. As the goaf

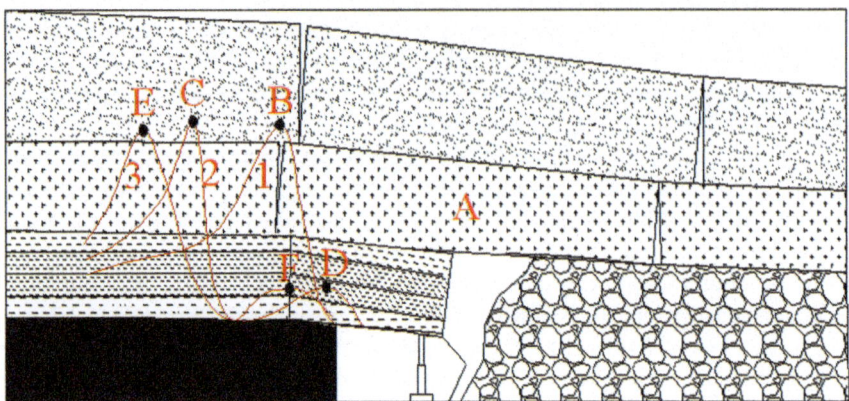

Fig. 2.15 The formation process of the internal stress field

continues to increase, the pressure transmitted to the coal wall through the roof rock beam is also constantly increasing, the peak of the mining stress external stress field continues to move to the outside, and the distribution range of the mining stress also continues to increase under a certain advancement distance. Curves 1, 2, and 3 show the process of the internal stress field going from existence to non-existence and the gradual transfer of the external stress field as the working face advances.

When the rock beam fractures again, a new internal stress field will be formed in front of the working face, and as the working face advances, the new internal stress field disappears, and there is such a process of formation and disappearance of the internal stress field in front of the working face.

C) The change rule of the internal stress field behind the cutting eye

Regarding the internal stress field behind the cutting eye, after its formation, because there is no coal wall advancement effect, therefore, without the upper rock beam fracturing again, the size and range of the internal stress field will not change significantly.

D) Derivation of the internal stress field

The basis for calculating the internal stress field is that the vertical mining stress distributed within the range of the internal stress field on the coal body around the goaf is equal to the weight transferred to this range after the fracture of the top rock layer of the "fracture arch". As shown in Fig. 2.16:

$$\frac{1}{2} S_0 \frac{K0_{\max}}{S_1} = \frac{H_g \cdot C_i \cdot \gamma}{2} \qquad (2.2)$$

where, S_0 is the range of the internal stress field, m; C_i is the basic top rock beam cycle pressure step distance, m; $K_{\max}$ is the stress concentration coefficient; H is the mining depth, m; S_1 is the distance from the peak position of the mining stress to the coal wall, m; γ is the rock layer bulk density, kN/m^3.

2.2 The Evolution and Development Law of Overburden and Mining Stress …

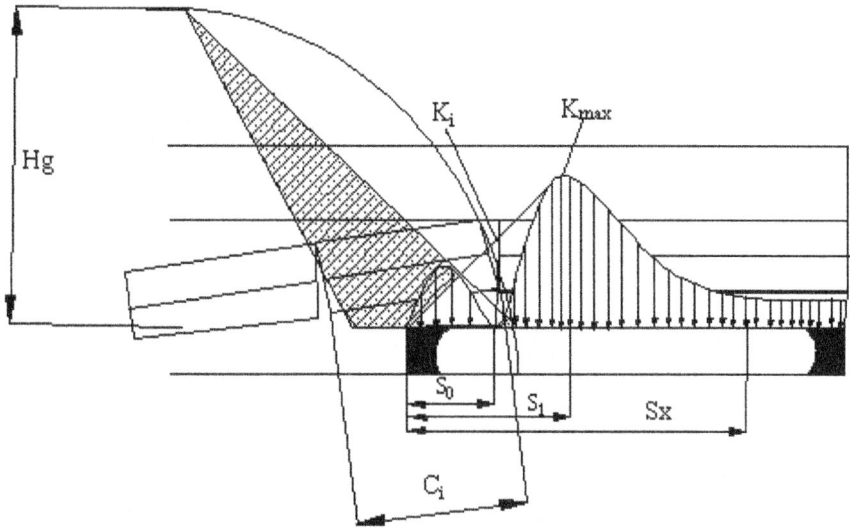

Fig. 2.16 Mining stress calculation model

The calculation formula 2.3 for the range of the internal stress field is obtained by solving Eq. 2.2,

$$S_0 = \sqrt{\frac{C_i \cdot H_g \cdot S_1}{K_{\max}}} \qquad (2.3)$$

When the basic top comes to pressure for the first time, $S_0 = \sqrt{\frac{0.5 C_i \cdot H_g \cdot S_1}{K_{\max}}}$.

3. The third stage—the relatively stable stage

The relatively stable stage of mining stress can be further subdivided into 2 stages.

A) The first stage of mining stress stability—relative stability in front of the mining field coal wall and behind the cutting eye

When the mining field advances to the inclined length of the mining field, according to the principle of Prussian natural balance arch, a relatively stable structural arch will form above the mining field. This structural arch transfers the weight of the rock above the arch to the foot of the arch through the arch line. Since the inclined length of the mining field is equal to the advance length at this moment, it is a semi-spherical structure in terms of spatial structure, as shown in Fig. 2.17. At this time, a uniform mining stress band is formed around the mining field. Without considering the influence of the advance speed, the A band and the B band will not change much with the advance of the mining field and tend to be stable. This stage of stability is called the first stage of mining stress stability.

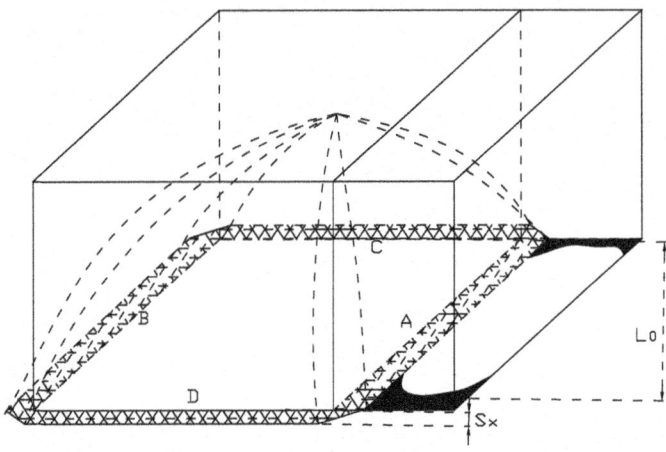

Fig. 2.17 The first stage of mining stress stability

For the C band and the D band, as the mining field advances, the A and B sides are pushed forward and backward respectively, and the two side bands will change, forming the second stability of mining stress.

B) The second stage of mining stress stability—the relative mining impact stability stage of the mining field boundary

When the mining field continues to advance, the foot of the balance arch has been transferred. At the front and rear ends of the mining field (i.e., A and B in Fig. 2.18), its original arch force structure is maintained. Therefore, the size of the A and B bands and the level of stress basically remain unchanged. But for the side bands, due to the increase in the distance between A and B, it cannot maintain the joint bearing effect on all overlying rock layers. The weight of the rock exceeding its bearing range will be transferred to the coal body on both sides of the mining field. That is, from the perspective of the development and change of mining stress, the mining stress on both sides of the mining field will develop in the direction of bands E and F, and finally reach its new stable state.

According to engineering practice and similar material simulation, when the mining field advances a distance of 0.75–1 times the working face length from this point, the mining field is basically stable, that is, in Fig. 2.18, the distance from point G to the coal mining working face is 0.75 times the inclined length of the mining field. Correspondingly, according to the analogy principle, the distance from point H to the cutting eye is also 0.75 times the inclined length of the mining field.

4. The development and change rule of mining stress of overlying rock in mining field

In actual coal mining work, mining stress generally goes through three development stages, but in strata where the coal seam is relatively shallow, the coal seam strength is relatively large, and the roof is relatively soft, there may not be the appearance of

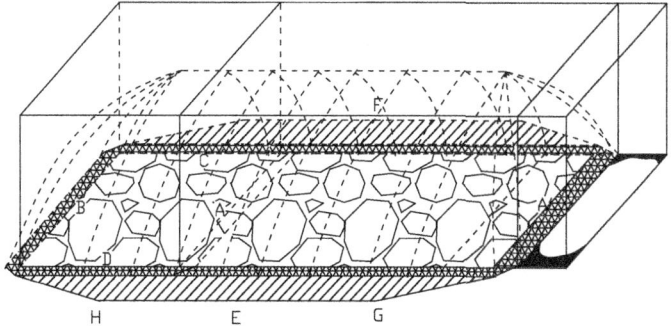

Fig. 2.18 The second stage of mining stress stability

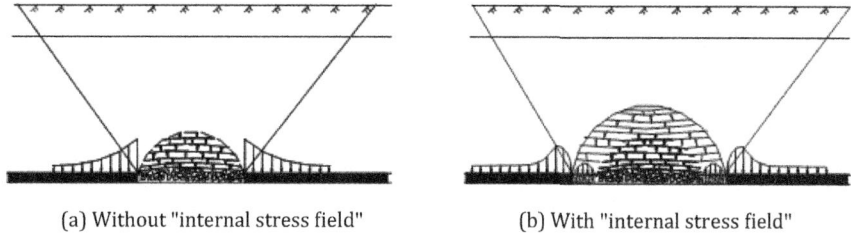

Fig. 2.19 Schematic diagram of mining stress distribution

the internal stress field in the second stage; or if the working face is reinforced, the roof rock beam may also break at the end of the support or at the coal wall. According to the research results of mining pressure in the mining field, there are two situations of mining stress state on the coal wall, as shown in Fig. 2.19.

A) Without "internal stress field", when the stress at each point on the coal wall has not reached the destruction limit of the coal body, the coal wall is in an elastic compression state, and the mining stress curve will be a monotonically decreasing curve with a peak at the coal wall.

B) With "internal stress field", when the stress value near the coal wall reaches the strength limit of the coal seam, with the destruction of the coal body, its bearing capacity decreases, and the pressure peak near the coal wall will transfer to the deep part of the coal body, dividing the mining stress distribution interval into elastic zone and plastic zone. The subsequent rock beam fracture will clearly divide the mining stress into two parts along the fracture line, namely the "internal stress field" (determined by the self-weight of the already fractured rock beam between the fracture line and the coal wall) and the "external stress field" (determined by the weight of the overlying rock load in the "stress arch" outside the fracture line).

At the same time, as known from the above content, after the coal seam is mined, the overlying rock produces fracture damage. This damage does not develop upwards indefinitely, but stops at a certain height, forming a "fracture arch". The rock beam above the "fracture arch" not only has a force connection with the rock beam inside the arch, but also has a force connection with the arch base at both ends. In the case of a small mining width, if there is a rock beam with a large stiffness above the arch, that is, the "support layer", then the deflection of this rock beam is very small. The rock beam above the "fracture arch" will transfer the force to the rock body above the coal wall on both sides. At this time, the rock beam inside the "fracture arch" is not affected by the pressure of the overlying rock layer, as shown in Fig. 2.20. If the stiffness of the "support layer" above the "fracture arch" is small, that is, the deflection of the rock beam is large, the working face will continue to advance after the mining area is squared. Under the action of the overlying rock layer, the "support layer" will gradually bend, and a part will settle on the rock beam inside the "fracture arch". At this time, the force will be divided into two parts, one part will be transferred to the rock body above the coal wall on both sides, and the other part will be borne by the rock beam inside the arch, as shown in Fig. 2.21.

The distribution of mining stress in the mining area is a three-dimensional structure, and the advancement speed of the working face has a certain impact on it. Especially under the background of increasing mining depth, working face support and mining intensity year by year, a scientific attitude should be applied to dialectically and dynamically view the distribution and evolution of mining stress under different mining area conditions.

Fig. 2.20 "Fracture arch" not affected by overlying rock layer pressure

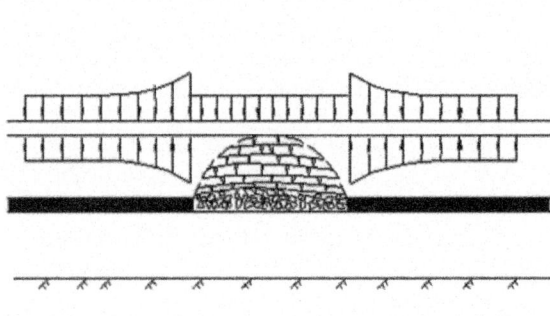

Fig. 2.21 "Fracture arch" affected by overlying rock layer pressure

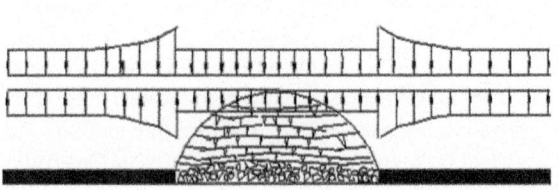

2.2.2 Distribution Pattern of Mining Stress on the Plane

The distribution law of mining stress on the plane (on coal and gangue) is the basis for controlling the mine pressure of the floor roadway. Under the guidance of the practical mine pressure theory system centered on "rock layer movement", analyze the stress evolution process in the mining area along the advancement direction and parallel to the working face direction.

1. Distribution law of mining stress along the advancement direction

During the normal advancement process, the distribution of mining stress in the advancement direction is as shown by the solid line in Fig. 2.22, and there is a leading pressure on the coal body in front of the working face. The pressure in the mined-out area behind the working face monotonically increases. After a certain distance behind the working face, the overlying rock activity stabilizes, and the pressure in the mined-out area rises to the original stress.

In addition to the direct top, each rock layer of the overlying rock layer in the mining area is supported by the coal body at one end and the gangue in the mined-out area at the other end, maintaining a force transmission connection in the advancement direction. During the normal advancement process, the overlying rock activity behind the working face develops from bottom to top. The upper rock layer either moves with the lower rock layer or lags behind the lower rock layer. The rock layers are in a delamination state or not all transmit rock weight. In the contact state, the rear support points of each group of rock layers from bottom to top are arranged one after another, so the pressure in the mined-out area behind the working face is monotonically increasing. Except for short working faces and large-depth mining geological conditions, the pressure in the mined-out area rises to the original stress after the overlying rock activity stabilizes behind the working face.

After the working face is mined, during the process of overlying rock activity stabilization, the overlying rock layers that move simultaneously with the basic roof decrease with the bending and settling span, and the support points in the mined-out area move towards the coal wall direction, or even further combine. The support points in the mined-out area overlap. During the process of overlying rock activity, the pressure on the outside of the basic roof rock beam touch point continuously rises. After stabilization, a pressure peak higher than the original stress will form on the outside of the basic roof rock beam touch point, while the pressure on the coal body in front drops to the lowest. At this time, the pressure distribution in the advancement direction is as shown by the dashed line in Fig. 2.22.

2. Distribution law of mining stress in the direction parallel to the working face

The distribution law of mining stress in the direction parallel to the working face will go through the following six stages:

A) Leading pressure stage: There is a large mining stress distributed on the coal body in front of the working face. For the working face with solid coal on both sides, from the groove to the middle of the working face, the roof is gradually

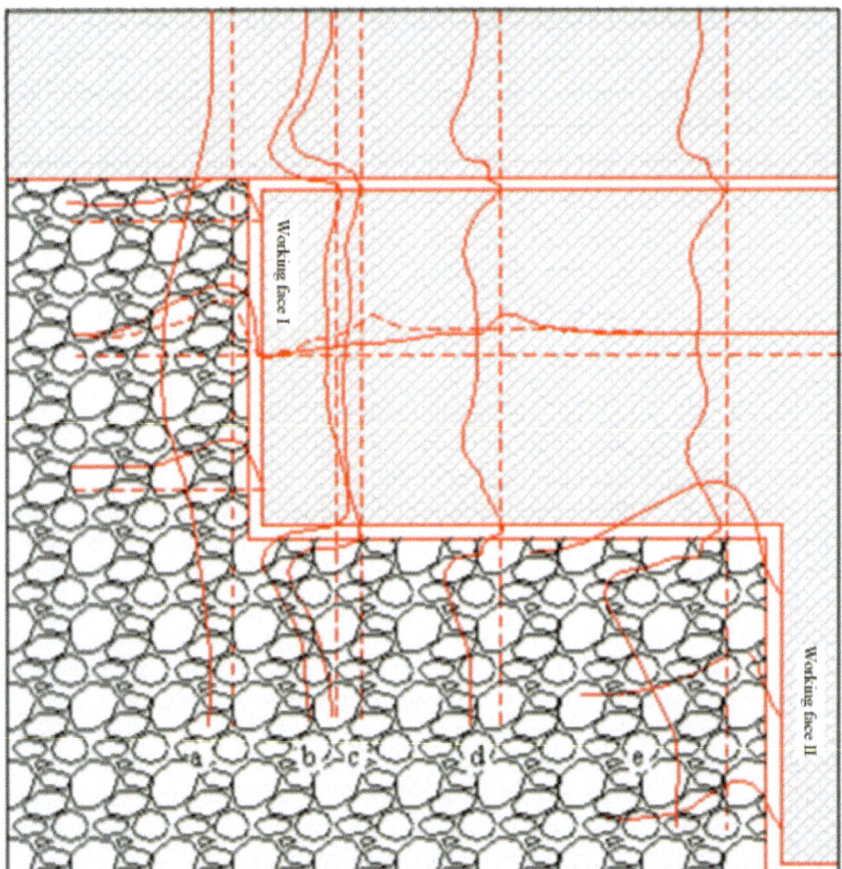

Fig. 2.22 Distribution law of mining stress on the plane

reduced by the support of the coal body on both sides, so the leading pressure gradually increases and reaches the maximum in the middle of the working face, as shown in 2.22. For the working face with one side mined out, the leading pressure increases from the solid coal end to the mined-out area end. On the convex corner of the coal body, due to the pressure overlap of the mined-out coal body on both sides, as shown in Fig. 2.22 e.

B) Relatively stable stage: From the coal wall of the working face to the fracture of the rock beam at the rear of the working face, the rock beam is in a relatively stable stage, and the pressure on the coal body shows a single peak curve; because the basic roof rock beam does not show significant movement, the goaf only bears the weight of the directly falling gangue, the pressure is very small, as shown in Fig. 2.22.

C) Significant movement stage: From the fracture of the basic roof rock beam to the completion of pressure. The distribution of mining stress on the coal body

2.2 The Evolution and Development Law of Overburden and Mining Stress ...

changes significantly with the development of significant movement of the basic roof rock beam, the main features are:

After the fracture of the rock beam is completed, the stress field is divided into two parts along the fracture line, that is, between the fracture line and the edge of the coal body, the "internal stress field" determined by the self-weight of the fractured rock beam, and the "external stress field" determined by the overall weight of the overlying rock layer outside the fracture line. At this time, the pressure distribution of the two stress fields is shown by the solid line in Fig. 2.22. When the basic roof beam sinks and touches the gangue: the "internal stress field" basically reaches stability, and the pressure distribution is shown by the dashed line in Fig. 2.22. After the basic roof rock beam touches the gangue, the pressure in the goaf increases.

It can be seen that the low stress area is formed after the rock beam is fractured, and it basically stabilizes after the rock beam rotates and touches the gangue. The range of the low stress area is from the fracture line of the rock beam to the edge of the coal body. The range and stability time of the low stress area are the basis for selecting the location and excavation time of the roadway.

D) Overburden stable stage: After the basic roof rock beam is pressurized, the overburden of the basic roof continues to fracture or bend and sink, touch the gangue, the stress of the external stress field continues to decrease, the internal stress slightly increases, and the pressure in the goaf continues to rise. When the overburden is stable, the stress of the external stress field drops to the lowest, the pressure in the goaf from the edge of the coal body to the direction of the goaf continues to increase, and a pressure peak higher than the original stress is formed on the outside of the touch point of the basic roof rock beam, as shown in Fig. 2.22.

E) Superimposed pressure stage: When approaching the advancement of the working face, the advanced pressure on the working face of the coal seam corner and the original mining stress are superimposed on each other. The stress of the internal and external stress fields will rise during this stage. Due to the specific geometric conditions at the corner of the coal seam, the stress concentration factor of the superimposed pressure peak can reach 5–10, as shown in Fig. 2.22e.

F) Pressure recovery stage: A certain distance behind the working face, after the overburden activity is stable, the pressure in the goaf returns to the original stress.

2.2.2.1 Analysis of Mining Stress Limit Equilibrium and Bearing Capacity

After the stope advances, the stress of the surrounding rock mass is redistributed, and the surrounding coal body is first damaged and gradually extends to the boundary of the elastic stress zone. The stress of this part of the coal body is in a state of stress limit equilibrium. Due to the Poisson's ratio of the coal body μ is greater than the Poisson's ratio of the rock of its roof and floor, the cohesion of the interface

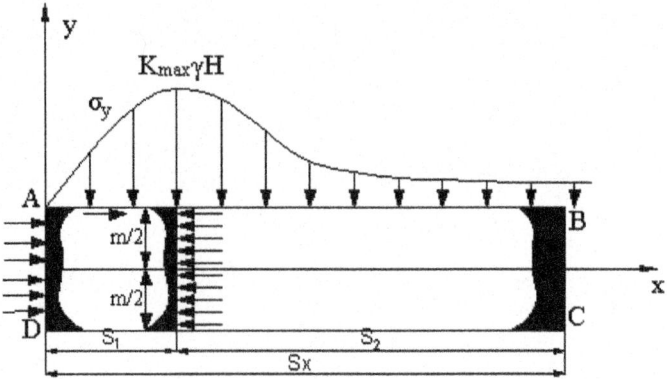

Fig. 2.23 Simplified diagram of mining stress calculation on coal seam interface

between the coal seam and the roof and floor rock C and the internal friction angle φ are lower than the values of the cohesion and internal friction angle of the coal body. After excavation, the coal body will inevitably be squeezed out from the rock of the roof and floor, and shear stress τ_{xy} is generated on the coal seam interface. The calculation mechanical model is shown in Fig. 2.23. In this figure, ABCD is the stress limit equilibrium zone; $\overline{\sigma_x}$ is the average value of the horizontal stress σ_x on the entire thickness of the coal wall at $= S_1$; P_x is the support resistance to the coal wall.

1. Basic assumptions
A) The coal seam interface is the slip surface of the coal body relative to the roof and floor rock layers. The normal stress σ_y and shear stress τ_{xy} on the slip surface should satisfy the stress limit equilibrium equation, that is:

$$\tau_{xy} = \sigma_y \cdot tg\varphi + C \tag{2.4}$$

B) Since the force of the goaf gangue on the coal wall is very small, it can be approximated as zero, that is, $P_x \approx 0$;
C) The stress of the coal body is symmetrical to the x axis.
D) At the junction of the stress limit equilibrium zone and the elastic zone (the junction of elasticity and plasticity), i.e., $x = S_1$ the equilibrium equation at this time is:

$$[\sigma_y]_{x=S_1} = K_{max}\overline{\sigma_x} = \lambda[\sigma_y]_{x=S_1} = \lambda K_{max} \tag{2.5}$$

2. Solution of theoretical model

The basic equation for solving the interface stress of the limit equilibrium zone is formula 2.6.

2.2 The Evolution and Development Law of Overburden and Mining Stress ...

$$\begin{cases} \dfrac{\partial \sigma_x}{\sigma_x} + \dfrac{\partial \tau_{xy}}{\sigma_y} = 0 \\ \dfrac{\partial \tau_{xy}}{\sigma_x} + \dfrac{\partial \sigma_y}{\sigma_y} = 0 \\ \tau_{xy} = \sigma_y \cdot tg\varphi + C \end{cases} \quad (2.6)$$

According to the mechanical model shown in the figure above, taking the coal body (ABCD) of the entire stress limit equilibrium zone as the separated body, the sum of the forces in the x direction is zero, the equilibrium equation can be obtained as formula 2.7

$$m \cdot \sigma_x - 2\int_0^{S_1} \tau_{xy} dx - p_x m = 0 \quad (2.7)$$

By solving Eqs. 2.5–2.7, the theoretical model of the mining stress σ_y and the width of the limit equilibrium zone S_1 of the coal body on the side of the mining area can be obtained:

$$\sigma_y = \frac{C}{f}\left(e^{\frac{2fx}{m\lambda}} - 1\right) S_1 = \frac{m\lambda}{2f} \ln\left(\frac{K_{\max}\lambda \cdot H \cdot f}{C} + 1\right) \quad (2.8)$$

where, σ_y positive stress on the coal body, MPa; σ_x horizontal stress, MPa; λ lateral pressure coefficient of the coal body, $\lambda = \frac{\mu}{1-\mu}$; C cohesion of coal seam and rock, MPa; φ internal friction angle of rock layer; f friction coefficient of coal seam interface, $f = tg\varphi$; $K_{\max}$ stress concentration factor; γ average bulk density of overlying rock layer, kN/m³; H burial depth of coal body, m; S_1 limit stress equilibrium zone (plastic zone), m; S_2 elastic zone, m; S_x range of mining stress influence, m.

3. Calculation of mining stress distribution range when the mining area is exposed

Through the previous analysis, it is known that the range of mining stress influence in front of the coal wall reaches its maximum when the mining area advances to the inclined length of the mining area. As the mining area advances, the range of mining stress influence in front no longer expands; a model as shown in Fig. 2.24 is established.

After the mining area advances to the exposure, under the action of gravity, the rock layer above the mining area forms a pressure increase belt with a width of S_x around the mining area. Ignoring the weight of the old pond gangue, Eq. 2.9 is established.

$$\left(2L_0 \cdot S_x + |2C_x|_{=L_0} \cdot S_x + 2 \cdot S_x^2\right) \cdot (K_a - 1) \cdot \gamma \cdot H = L_0| \cdot C_x|_{=L_0} \cdot \gamma \cdot H \quad (2.9)$$

where, K_a is the average value of the stress concentration factor.
Simplify Eq. 2.9 to get

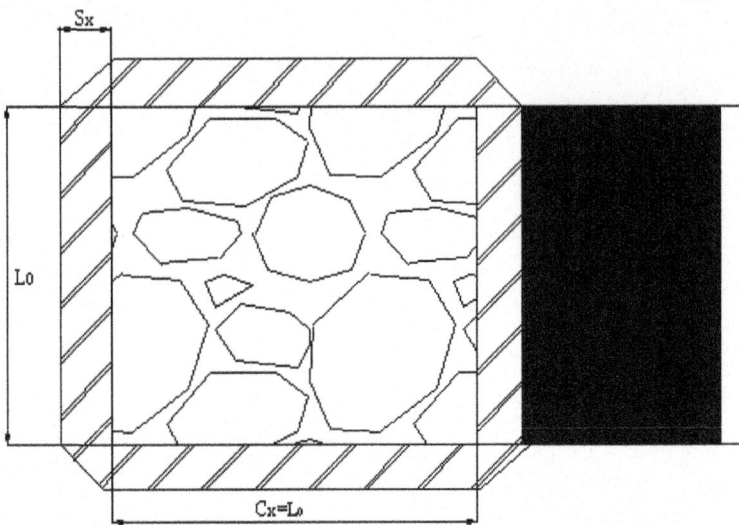

Fig. 2.24 Schematic diagram of the distribution range of mining stress when the mining area is exposed

$$S_x^2 + 2L_0 \cdot S_x - \frac{L_0^2}{2(K_a - 1)} = 0$$

Solve the equation to get $S_x = \frac{-2L_0 \pm \sqrt{4L_0^2 + 4\frac{L_0^2}{2(K_a-1)}}}{2}$ discard the negative root to get

$$S_x = L_0\left(\sqrt{1 + \frac{1}{(2K_a-2)}} - 1\right) \quad (2.10)$$

It can be seen from formula 2.10 that the distribution range of mining stress is related to the length of the inclined direction of the mining area and the average concentration factor of mining stress.

Considering the influence of the mining area inclination angle, revise formula 2.10 as follows:

$$\begin{aligned} S_{x1} &= L_0\left(\sqrt{1 + \frac{1}{(2K_a-2)}} - 1\right)(1 + \sin\alpha) \\ S_{x2} &= L_0\left(\sqrt{1 + \frac{1}{(2K_a-2)}} - 1\right)(1 - \sin\alpha) \\ S_{x3} &= L_0\left(\sqrt{1 + \frac{1}{(2K_a-2)}} - 1\right) = S_{x4} \end{aligned} \quad (2.11)$$

where, S_{x1} represents the distribution range of mining stress in the downhill direction of the mining area; S_{x2} represents the distribution range of mining stress in the uphill direction of the mining area; S_{x3} represents the distribution range of mining stress in front of the coal wall of the mining area; S_{x4} represents the distribution range of mining stress behind the coal wall of the mining area.

4. Influence law of mining stress during mining area advancement

When the advancement length of the mining field is equal to the inclined length, the fracture arch above the working face reaches its maximum, as shown in Fig. 2.24. At this time, the weight of the overlying rock layer is evenly distributed around the mining field. Through practice, it is known that when the working face continues to advance, the range of mining stress distribution between the front of the mining field and the back of the cutting eye does not change much, but the weight of the overlying rock layer supported by the two sides of the mining field will become larger and larger due to the distance between the coal wall in front of the mining field and the coal wall at the back, until the coal body per unit length on both sides of the mining field bears all the weight of the overlying rock layer within the unit length, as shown in Fig. 2.25.

A) When the stiffness of the "support layer" is large and the load of the overlying rock layer on the "fracture arch" is borne by the coal body on both sides

$$(2L_A \cdot S_{x(\max)}) \cdot (K_a' - 1) \cdot \gamma \cdot H = L_0 \cdot L_A \cdot \gamma \cdot H - \tfrac{1}{2}\pi\left(\tfrac{1}{2}L_0\right) \cdot H_g \cdot \gamma_g \cdot L_A$$

$$S_{x(\max)} = \frac{L_0(4H - \pi H_g)}{8(K_a' - 1)H} \tag{2.12}$$

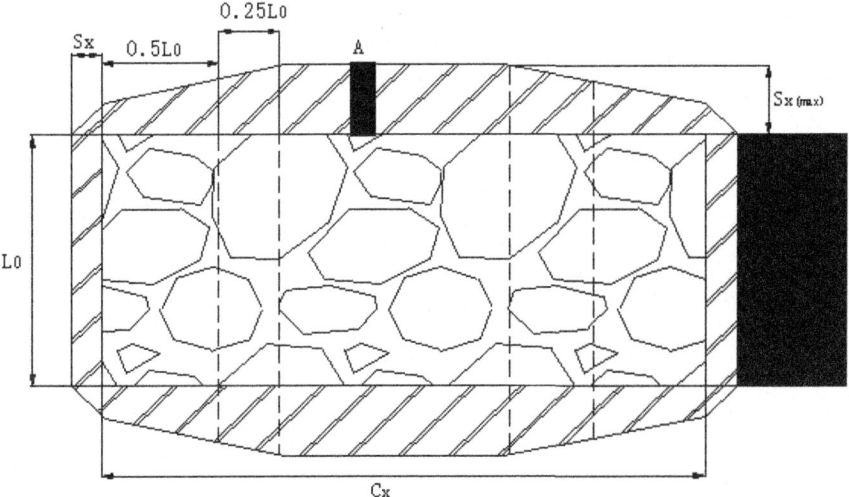

Fig. 2.25 Schematic diagram of the stable state behind the mining stress distribution range

where, K'_a is the maximum average value of the stress concentration coefficient; $S_{x(\max)}$ is the maximum value of the mining stress distribution range, m; H_g is the height of the structural arch, m.

B) When the stiffness of the "support layer" is small and part of the load of the overlying rock layer acts on the rock beam in the arch

$$(2L_A \cdot S_{x(\max)}) \cdot (K'_a - 1) \cdot \gamma \cdot H = L_0 \cdot L_A \cdot \gamma \cdot H - Q^1_{arch} Q^1_{arch} > \frac{1}{2}\pi \left(\frac{1}{2}L_0\right) \cdot H_g \cdot \gamma_g \cdot L_A \quad (2.13)$$

At this time, the position of the stress peak will decrease. Where, $Q = \frac{L_0 H_g}{2\gamma H}$ is a constant, K'_a is the maximum average value of the stress concentration coefficient; $S_{x(\max)}$ is the maximum value of the mining stress distribution range.

2.2.3 Relationship Between Overburden Failure Structure and Mining Stress Evolution

The stress of the roadway surrounding rock excavated and maintained in the internal stress field comes from the force of the overlying rock layer that moves significantly under the influence of mining. As the mining field advances into the significantly moving rock layer, including the directly collapsed roof (M_Z) and the basic roof that maintains the transmission force connection in motion ($M_{E1}, M_{E2}...M_{EN}$). The basic structural state and structural parameters of the movement are shown in Fig. 2.26: The pressure of the rock layer movement within this range on the coal body in the internal stress field can be based on the following two situations that may occur when the rock beam (plate) under the basic roof moves:

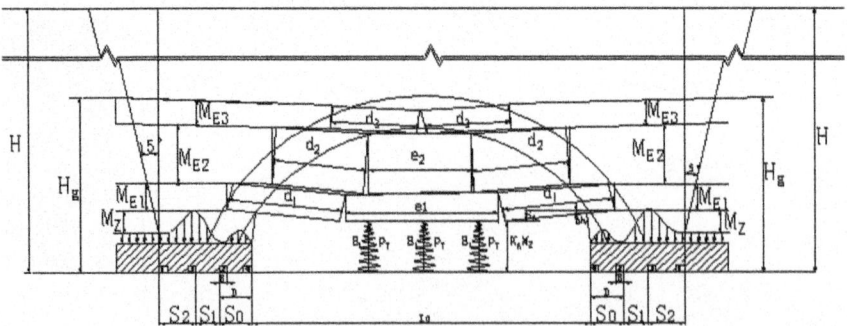

Fig. 2.26 Stress sources of roadway surrounding rock at different positions **a** Original stress field; **b** Internal stress field; **c** External stress field

2.2 The Evolution and Development Law of Overburden and Mining Stress ...

The first situation: the shear instability at the end of the rock beam, that is, the bite point 0 in Fig. 2.26 loses the squeezing and twisting ability.

Under this situation, the pressure on the compressed coal body (S_0) in the internal stress field P_{S0} and the vertical stress in the coal seam σ_S can be obtained according to the gravity balance program of the basic roof rock beam. Among them: when the first rock beam (lower rock beam) fractures, its motion gravity balance equation is:

$$2 \cdot \int_0^{S_0} \sigma_S ds + E_i \varepsilon_i l_1 = 2A + B \tag{2.14}$$

That is:

$$P_{S0} = \int_0^{S_0} \sigma_S ds = \tfrac{1}{2}[(2A + B_1) - E_i \varepsilon_i l_1] \tag{2.15}$$

where, $A = m_Z \cdot \gamma_Z \cdot l_Z$, is the direct roof force; $B_1 = (L_0 + 2S_0)m_{E1} \cdot \gamma_{E1}$, is the basic roof force; $P_T = E_i \varepsilon_i e_1$, is the support reaction force of the collapsed rubble; M_Z and γ_Z are the thickness and bulk density of the direct roof (collapsed rock layer) respectively; M_{E1} and γ_{E1} are the thickness and bulk density of the first (lower) rock beam of the basic roof respectively. L_Z and L_0 are the span of the direct roof and the length of the working face respectively.

The related symbol meanings of the rubble reaction force P_T value are: e_1 is the span of the middle section of the basic roof lower rock beam shear fracture (m); ε_i is the settlement amount of the basic roof rock beam after touching the rubble (i.e., the compression amount of the old pond rubble) (mm); E_i For the compressive stiffness of the caving gangue (tons/mm· m^2); where, $e_1 = (L_0 + 2S_0) - (d_1 + d_2)$.

For nearly horizontal coal seams, studies have shown that when the first fracture step of the basic roof lower rock beam is known, the following relationship holds, that is:

$$d_1 = d_2 \approx C_{01}$$

The settlement of the basic roof lower rock beam can be expressed by the following formula:

$$\varepsilon_i = S_i - S_A$$

where, S_i is the settlement value of the basic roof lower rock beam (mm). S_A is the settlement value of the basic roof lower rock beam at the time of contact (mm), which can be obtained by the following formula, that is:

$$S_A = h - m_Z(K_A - 1)$$

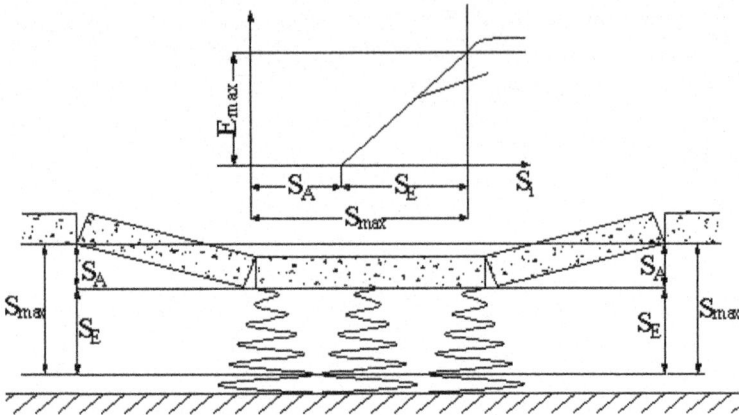

Fig. 2.27 Relationship diagram of swelling gangue compressive stiffness

where, h is the mining height, m_Z is the thickness of the direct roof (caving rock layer), K_A is the swelling coefficient of the caving rock layer before compression. Generally, K_A is taken as 1.35–1.45.

Experimental studies have shown that the relationship curve of the compressive stiffness of the swelling gangue $E_i = f(S_i)$ is shown in Fig. 2.27.

The compressive stiffness equation of the approximate linear segment can be expressed by the following formula:

$$E_i = f(S_i) = \frac{E_{max}}{SA_{Amax}(S_i - S_A)} \tag{2.16}$$

where, S_{max} is the maximum settlement value of the rock beam when the swelling gangue at the old pond caving is in the "limit compression" state (i.e. $E_i = E_{max}$). It can be expressed by the following formula:

$$S_{max} = h - m_Z(K_{min} - 1)$$

K_{min} is the swelling coefficient of the caving rock layer when it is in the "limit compression" state. From a practical point of view, it can be taken between K_{min}. From this, the expression for the support reaction P_T value during the rock beam settlement process can be obtained:

$$P_T = E_i \varepsilon_i e_1 = \frac{E_{max}}{SA_{max}(S_i - S_A)(S_i - S_A)e_1} = \frac{e_1 E_{max}}{SA_{max}(S_i - S_A)^2}$$

The expression for this is:

$$P_T = C(S_i - S_A)^2 \tag{2.17}$$

where, $= \frac{e_1 E_{max}}{SA_{max}}$.

Substituting the listed results into formula 2.15, the pressure equation of the basic roof lower rock beam fracture action on the internal stress field coal body—the expression of the "rock beam position equation" is obtained:

$$P_{S0} = \int_0^{S_0} \sigma_S ds = \frac{1}{2}\left[(2A + B_1) - C(S_i - S_A)^2\right] \tag{2.18}$$

If it is assumed that the stress in the compressed coal body of the internal stress field is uniformly distributed, then:

$$P_{S0} = \sigma_{sp} \cdot S_0 = \frac{1}{2}\left[(2A + B_1) - C(S_i - S_A)^2\right]$$

From this, the expression for the average vertical compressive stress on the coal body of the internal stress field is obtained:

$$\sigma_{sp} = \frac{1}{2S_0}\left[(2A + B) - C(S_i - S_A)^2\right] \tag{2.19}$$

The characteristics of this position equation are shown in Fig. 2.28.
When $S_i = 0 \to S_i = S_A$, $P_{S0} = P\frac{B}{2}_{Smax}$.
When $S_i = S_{max1}$, P_{Smin}.

S_{max1} is the settlement value at the touch gangue (point A) when the lower rock beam moves alone to enter the "final" stable state. When the rock beam settles to this position, the coal seam in the internal stress field only bears the weight of the "direct roof", that is:

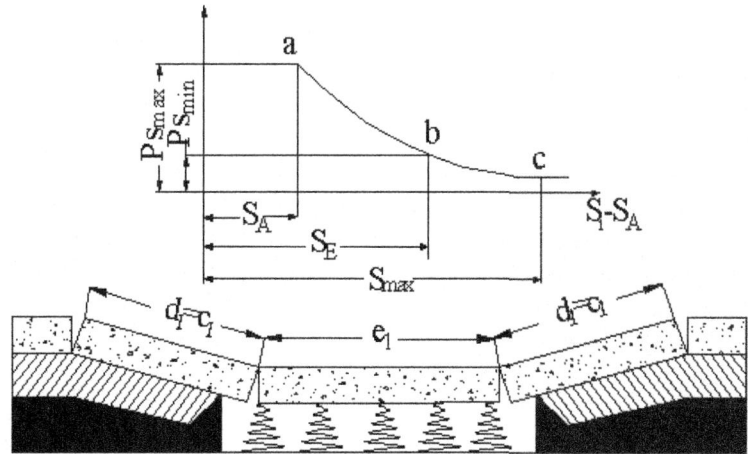

Fig. 2.28 Basic roof rock beam position equation diagram

$$P_{S0} = \frac{1}{2}[(2A + B_1) - C(S_i - S_A)^2] = A$$

From this, the value of S_{max1} can be obtained:

$$S_{max1} = \sqrt{\frac{B_1}{C}} + S_A \qquad (2.20)$$

where, $= m_{E1} \cdot \gamma_{E1} \cdot (L_0 + 2S_0)$; $C = \frac{e_1 E_{max}}{S_{max}}$; $S_A = h - m_Z \cdot (K_A - 1)$.

By the same token, we can derive the pressure position equation on the coal body of the internal stress field when each rock beam in the damaged arch is pressed to sink, and the final sinking value of the corresponding lower rock beam at the energy waste point (A). Among them:

When the second rock beam is broken and pressed, the following applies:

$$P_{S02} = \int_0^{S_0} \sigma_s ds = \frac{1}{2}[(2A + B_1 + B_2) - C(S_i - S_A)^2]$$

$$\sigma_{sp2} = \frac{1}{2S_0}[(2A + B_1 + B_2) - C(S_i - S_A)^2]$$

$$S_{max2} = \sqrt{\frac{B_1 + B_2}{C}} + S_A$$

When the n rock beam is broken and pressed, the following applies:

$$P_{S0n} = \int_0^{S_0} \sigma_s ds = \frac{1}{2}\left[\left(2A + \sum_1^n B_i\right) - C(S_i - S_A)^2\right] \qquad (2.21)$$

$$\sigma_{spn} = \frac{1}{2S_0}\left[\left(2A + \sum_1^n B_i\right) - C(S_i - S_A)^2\right] \qquad (2.22)$$

$$S_{maxzn} = \sqrt{\frac{\sum_1^n B_n}{C}} + S_A \qquad (2.23)$$

When all rock layers in the damaged arch range are broken and moved, the overall movement position equation of the force source of the internal stress field and the approximate expression of the final settlement value of the lower rock beam at the touch waste point are as follows:

$$P_S = \int_0^{S_0} \sigma_s ds = \frac{1}{2}[(2A + B_s) - C(S_i - S_A)^2] \qquad (2.24)$$

$$\sigma_{sp} = \frac{1}{2S}[(2A + B_S) - C(S_i - S_A)^2] \qquad (2.25)$$

$$S_{maxS} = \sqrt{\frac{B_S}{C}} + S_A \approx S_{max} \qquad (2.26)$$

In the formula,

$$B_S \approx \left[\Pi\left(\frac{L_0+2S_0}{2}\right)^2 - (h+m_Z)L_0\right]\gamma_P \tag{2.27}$$

$$S_{\max} = h - m_Z(K_{\min} - 1) \tag{2.28}$$

In summary, it can be clearly seen from the study that the mining stress acting on the coal body of the internal stress field is a function of the settlement of the overlying rock layer. Therefore, while determining the range of the internal stress field (S_0), it is important to understand the development of the movement of the overlying rock layer, and on this basis, choose the location and timing of the tunnel excavation correctly, and try to excavate and maintain the tunnel in a stable internal stress field, which is the key to controlling the mine pressure manifestation of the internal stress field tunnel.

The second situation: During the entire process of the movement of the overlying rock layer in the damaged arch, the lower rock beam at the end, that is, point 0 in Fig. 2.26, always maintains the transmission force connection.

Under this situation, the structural mechanics model of the movement process of the basic top lower rock beam is shown in Fig. 2.26. At this time, the pressure borne by the coal layer on the internal stress field should accept the movement of the rock beam, and the force moment balance equation for the bite point 0 is derived.

$$P_{S0} = \int_0^{S_0} \sigma_S ds = A + B \tag{2.29}$$

In the formula, A is the pressure given to the coal layer by the direct top movement; B is the pressure given to the coal layer by the movement of the basic top lower rock beam.

The pressure A given to the coal layer by the direct top movement (including its own movement and the forced movement under the basic top lower rock beam) can be obtained by taking the moment of the weight of the direct top at point 0:

$$A = \int_0^{S_0} \sigma_{SA} \cdot ds = m_Z \cdot \gamma_Z \cdot \frac{l_Z^2}{S_0} = m_Z \cdot \gamma_Z \cdot f_Z \tag{2.30}$$

In the formula, $f_Z = \frac{l_Z^2}{2l_S}$ is the moment coefficient; m_Z, L_Z, γ_Z are the thickness, span and bulk density of the direct top respectively. L_S is the distance from the point of action of the resultant force of the mining stress of the coal layer to the beam end (point 0).

$$P_{SOA} = \int_0^{S_0} \sigma_{SA} ds = \sigma_A \cdot S_0$$

$$l_S = \frac{S_0}{2}$$

$$fz = \frac{l_Z^2}{2\frac{S_0}{2}} = \frac{l_Z^2}{S_0}$$

From this, the pressure and average stress of the direct top are as follows:

$$A = m_Z \cdot \gamma_z \cdot \frac{l_Z^2}{S_0} \tag{2.31}$$

If $l_Z = S_0$, then:

$$A = m_Z \cdot \gamma_z \cdot S_0$$
$$\sigma_{AP} = m_Z \cdot \gamma_z \tag{2.32}$$

The force moment balance equation for the movement of the basic top lower rock beam is:

$$\int_0^{S_0} \sigma_{SB} ds \cdot l_S + \frac{d_1}{2} E_i \varepsilon_{1i} e_1 = \frac{1}{2} m_{E1} \cdot \gamma_{E1} \cdot e_1 d_1 + \frac{1}{2} m_{E1} \cdot \gamma_{E1} \cdot d_1 d_1 \tag{2.33}$$

From this, the pressure B1 acting on the coal layer of the internal stress field by the movement of the basic top lower rock beam is expressed by the following formula:

$$B_1 = \int_0^{S_0} \sigma_{SB} ds = \frac{m_{E1} \cdot \gamma_{E1} \cdot d_1}{2l_S}(e_1 + d_1) - \frac{d_1}{2l_S} E_i \varepsilon_i e_i \tag{2.34}$$

If it is assumed that the mining stress in the internal stress field is uniformly distributed, then the mining stress and average stress on the coal layer can be obtained as follows:

$$B_{P1} = \frac{m_{E1} \cdot \gamma_{E1} \cdot d_1}{S_0}(e_1 + d_1) - \frac{d_1}{S_0} E_i \varepsilon_{li} e_i \tag{2.35}$$

$$\sigma_{BP1} = \frac{m_{E1} \cdot \gamma_{E1} \cdot d_1}{S_0^2}(e_1 + d_1) - \frac{d_1}{S_0^2} E_{li} \varepsilon_i e_1 \tag{2.36}$$

In the formula, M_{E1}, γ_{E1} respectively for the thickness and bulk density of the basic roof lower rock beam; d_1 is the collapse degree of the rear block segment after the lower rock beam is fractured, studies have shown that under general roof conditions, it is close to the pressure step distance of this rock beam, that is:

2.2 The Evolution and Development Law of Overburden and Mining Stress …

$$d_1 \approx C_{10}$$

e_1 is the span of the middle bad segment after the lower rock beam is fractured, for nearly horizontal coal seams.

$$e_1 = (L_0 + 2S_0) - 2d_1$$

S_0 is the range of the inteal stress field; ε_{li} is the compression amount of the old pond fallen gangue from the start of the lower rock beam settlement energy to the entry into a stable state, which can be obtained by the following formula:

$$\varepsilon_{1i} = SA_{max}$$

In the formula,

$$S_{max} = h - m_Z(K_A - 1)$$
$$S_{max} = h - m_Z(K_{min} - 1)$$

In the formula, S_A is the settlement value when the lower rock beam starts to be able; S_{max} is the settlement value of the rock beam when the lower rock beam enters a stable state; K_A is the swelling coefficient of the compression part of the old pond fallen gangue; K_{min} is the swelling coefficient of the old pond fallen gangue after the lower rock beam enters a stable state; h is the mining height.

E_i is the compression stiffness of the old pond fallen gangue. As mentioned before, it has the characteristic of approximately linear increase with the compression increment. The increment law of the lower rock beam from touching the gangue to entering the final stable state, that is, $S_{1i} = S_{1max}$ can be expressed by the following formula, that is:

$$E \frac{E_{max}}{SA_{max}(S_i - S_A)_i} \tag{2.37}$$

Substitute the above analysis and research results into formula 2.28, after sorting, the pressure relationship equation of the internal stress field under the pressure of the lower rock beam fracture can be obtained—the "position equation" of the lower rock beam is:

$$P_{SP_1} = (A_1 + D_1) - C(S_i - S_A)^2 \tag{2.38}$$

In the formula,

$$A_1 = m_Z \cdot \gamma_Z \cdot \frac{l_Z^2}{S_0}$$

$$D_1 = \frac{m_{E1}\gamma_{E1}d_1(e_1 + d_1)}{S_0}$$

$$C = \frac{d_1 e_1 E_{max}}{S_0(S_{max} - S_A)}$$

If $S_{P1} = A$, the "final" settlement value S1max at the touch gangue under the separate action of the lower rock beam can also be calculated as:

$$S_{1\,max} = \sqrt{\frac{D_1}{C}} + S_A \qquad (2.39)$$

Based on the force position equation of the internal stress field coal body under the single movement of the lower rock beam, it is not difficult to derive the expression of the overall position equation of the overlying rock layer in the mining area according to the force moment balance condition:

$$P_{SP} = (A + D_n) - C(S_i - S_A)^2 \qquad (2.40)$$

In the formula,

$$D_n = \frac{\sum_1^n m_i \gamma_E d_i (e_i + d_i)}{S_0}$$

The position equation expression of the average pressure stress (σ_{SP}) is:

$$\sigma_{SP} = \frac{P_{SP}}{S_0} = \frac{1}{S_0}\left[(A + D_n) - C(S_i - S_A)^2\right] \qquad (2.41)$$

The final settlement value at the energy gangue of the lower rock beam when it is "finally" stable (S_{max}) is:

$$S_{n\,max} = \sqrt{\frac{D_n}{C}} + S_A \qquad (2.42)$$

Field practice has proved that when the working face length is within the limit of about 200 m, when the working face advances beyond the working face length, the rock layer with obvious movement in the overlying rock layer of the mining area, including the direct roof (fallen rock layer) and the basic roof rock beams ("fracture arch" inner rock layer) first fracture movement, is basically realized. After a period of settlement and compression of the fallen gangue, it can enter a relatively stable state.

2.2 The Evolution and Development Law of Overburden and Mining Stress ...

At this time, the settlement value at the touch gangue of the lower rock beam can be regarded as reaching the balance limit corresponding to the settlement force of the obviously moving overlying rock layer. At the same time, the compression stiffness (E) of the fallen gangue also reached the maximum value (Emax) corresponding to the force of the moving overlying rock layer. Comprehensive results of Chongzhou top coal mining and some other deep mine mining practices. For the mining area with a working face length of no more than about 200 m, the final settlement value at the energy gangue of the lower rock beam when the first movement of the overlying rock layer of the mining area is completed can be estimated by the following formula.

$$S_{max} = h - m_z(K_{min} - 1) \tag{2.43}$$

In the formula, h is the mining height, m_Z For the direct top (caving stratum) thickness.

K_{min} When the lower rock beam enters a stable state, the minimum swelling coefficient after the old pond caving gangue is compacted. According to relevant research practices, from a practical point of view, take K_{min}. The results obtained can be substituted into formula 2.41 to obtain the maximum compressive stiffness value at the old pond caving gangue, even if

$$S_{nmax} = \sqrt{\frac{D_n}{C}} + S_A = S_{max}$$

$$\sqrt{\frac{D_n}{C}} = S_{max} - S_A$$

$$C = \frac{D_n}{(S_{max} - S_A)^2}$$

$$C = \frac{d_1 e_1 E_{max}}{S_0(S_{max} - S_A)}$$

The maximum compressive stiffness of the old pond caving gangue can be approximately obtained.

$$E \frac{D_n S_0}{d_1 e_1 (S_{max} - S_A)} \tag{2.44}$$

In the formula,

$$D_n = \frac{\sum_1^n m_{Ei} \gamma_{Ei} d_{Ei}(e_{Ei} + d_{Ei})}{S_0}$$

$$S_A = h - m_Z(K_A - 1)$$

$$S_{max} = h - m_Z(K_{min} - 1)$$

During the whole process of basic roof fracture and subsidence, the coal seam within the range of the internal stress field (S_0) will be compressed under the pressure of the corresponding rock beam, and its compression amount will increase with the increase of the subsidence amount of the lower rock beam S_i, until the lower rock beam enters a "relatively stable" state. Among them, when the lower rock beam moves alone, the maximum compressive deformation on the coal seam of the internal stress field is speculated by using the gravity balance position equation:

$$\Delta h_{1\,max} = \Delta h_{1A} + \Delta A_{1\varepsilon} = \frac{S_0 S_{1\,max}}{d_1} \tag{2.45}$$

In the formula,

$$S_{1\,max} = \sqrt{\frac{B_1}{C}} + S_A$$

Substitute to get:

$$\Delta h_{1\,max} = \frac{S_0}{d_1}\left(S_A + \sqrt{\frac{B_1}{C}}\right) \tag{2.46}$$

Similarly, the force moment balance position equation can be used to guide the calculation:

$$\Delta h_{i\,max} = \frac{S_0}{d_1}\left(S_A + \sqrt{\frac{D_1}{C}}\right) \tag{2.47}$$

When all the rock beams of the basic roof fracture and move, the maximum compression amount of the coal body in the internal stress field is approximated by the gravity balance force moment balance position equation:

$$\Delta h_{max} = \frac{S_0}{d_1}\left(S_A + \sqrt{\frac{\sum_{1}^{n} B_1}{C}}\right) \tag{2.48}$$

$$\Delta h_{max} = \frac{S_0}{d_1}\left(S_A + \sqrt{\frac{D_n}{C}}\right) \tag{2.49}$$

Under the action of structural stress, the top plate buckling failure will develop layer by layer from bottom to top. Until due to the reduction of the span, the rock layer that can maintain stability under axial pressure. When the roadway roof is composed of rock layers with not large stratified thickness and not high strength. Given that the span of each rock layer gradually decreases from bottom to top. Therefore, the range of roof failure will generally be arch-shaped, and the roof pressure value can be calculated as a semicircle for support design.

For the original stress field with structural stress, the roadway roof failure is fully realized, and after the release of structural stress, the self-weight stress of the overlying rock layer will play a role. The two sides of the roadway failure under gravity will follow the failure law of a single gravity field, and the corresponding support design can be carried out in the same procedure.

In the excavation and maintenance of the roadway in the mining face forming the mining stress peak area, the key to the control and support design of the roadway roof is also to first correctly determine the depth of the two sides of the failure and the corresponding range of the roof failure. Based on this, the roadway support design is carried out. The key to controlling mine pressure manifestation (including the control of surrounding rock deformation and support impedance force and deformation) in the excavation and maintenance of the roadway in the internal stress field includes the following three aspects: First, choose the location of the roadway excavation reasonably. Obviously, based on the determined range of the internal stress field, that is, the width of the coal belt that has entered the damage (So), try to excavate the roadway directly at the deep boundary of the internal stress field, that is, the boundary between the internal and external stress fields. That is to say, if the recovery rate is not considered, when excavating the roadway in the internal stress field, it is better to leave the coal pillar width larger, so as to reduce the pressure and corresponding deformation on the protective coal pillar to the minimum. Of course, if the roadway is excavated after the internal stress field is completely stable, as long as it is in the internal stress field, the width of the remaining roadway coal pillar only needs to ensure that there is no old pond leakage, and a little bit is not a problem [11, 12]. Secondly, correctly determine the time of tunnel excavation. Ensure that the tunnel is excavated in the stress field after the mining face has advanced a certain distance and time, and always keep the lagging distance and time within the target of excavating and maintaining the tunnel in a stable stress field, which is the key to controlling tunnel deformation and damage, and is of paramount importance. This point has been clearly demonstrated from the research on the pressure (stress) and deformation of the stress field coal seam strip, including the relationship expressed by the related displacement equation. Finally, according to the selected tunnel excavation maintenance time and the possible "stress field" force deformation development process, carry out the correct (targeted) support design.

As the mining field advances, the range of the stress field will appear on both sides, and the pressure and corresponding compression deformation process that the coal seam bears, as shown in Fig. 2.29.

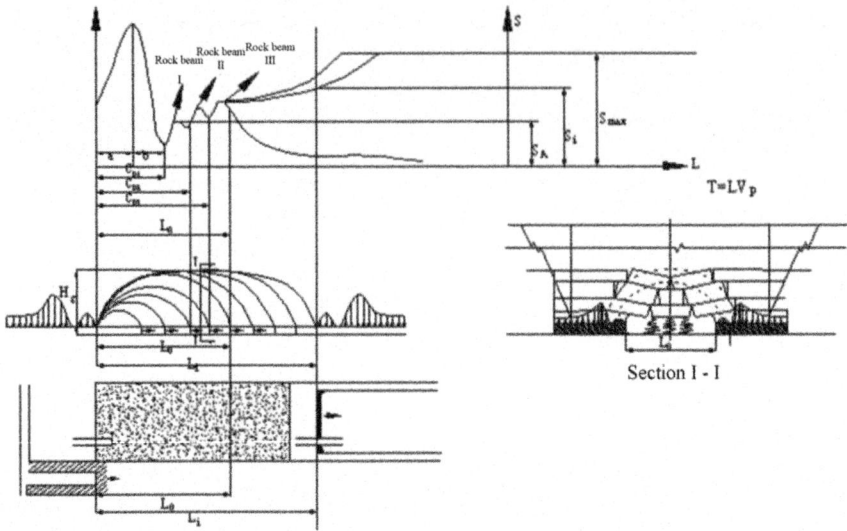

Fig. 2.29 Mining field advancement "stress field" range coal seam pressure and compression process

References

1. Qian M, Shi P, Xu J (2010) Mine pressure and strata control. China University of Mining and Technology Press
2. Zhang W (2008) Research on the structural mechanics model of coal mine mining field and its application. Shandong University of Science and Technology, Qingdao
3. Ma Q (2005) Research on "O" type spatial structure of overburden in longwall mining field and related mine pressure. Shandong University of Science and Technology, Qingdao
4. Yang W, Lin B, Qu Y et al (2010) Numerical simulation study on spatial distribution of stress field induced by coal seam mining. Chen B, Li G (eds) Proceedings of 2010 (Shenyang) international conference on safety science and technology. Northeast University Pres, Shenyang, pp 339–342
5. Jiang F (2006) Research on the concept of spatial structure of overburden in mining field and its application. J Min Safety Eng (01):30–33
6. Wei H, Jiang F, Wang C et al (2009) "C" type spatial structure of overburden in three-side mining field and its pressure control. J Coal 34(03):310–314
7. Guo W (2012) Research on the deformation evolution law of overburden in deep mining. Coal Industry Publishing House
8. Cao H (2019) Study on the three-dimensional stress change law of coal mine roof rock layer under mining influence. J Water Resour Archit Eng 17(04):112–116
9. Lin Y, Tu M, Fu B et al (2019). Research on the mechanical mechanism and control of fault stability under mining influence. Coal Sci Technol 47(09):158–165
10. Han G, Li X, Qu X et al (2019) Research on the correlation between spatial rupture of overlying strata in mining field and distribution of mining stress field. Coal Sci Technol 47(02):53–58
11. Zhang Y, Wan Z, Li F et al (2012) Mechanism of large deformation of surrounding rock under unstable overburden along empty excavation roadway. J Min Safety Eng 29(04):451–458
12. Jia B (2011) Deformation and destruction of roadway along empty space under mining influence and anchor support. J Liaoning Tech Univ (Natural Science Edition) 30(06):810–813

Open Access This chapter is licensed under the terms of the Creative Commons Attribution-NonCommercial-NoDerivatives 4.0 International License (http://creativecommons.org/licenses/by-nc-nd/4.0/), which permits any noncommercial use, sharing, distribution and reproduction in any medium or format, as long as you give appropriate credit to the original author(s) and the source, provide a link to the Creative Commons license and indicate if you modified the licensed material. You do not have permission under this license to share adapted material derived from this chapter or parts of it.

The images or other third party material in this chapter are included in the chapter's Creative Commons license, unless indicated otherwise in a credit line to the material. If material is not included in the chapter's Creative Commons license and your intended use is not permitted by statutory regulation or exceeds the permitted use, you will need to obtain permission directly from the copyright holder.

Chapter 3
Temporal and Spatial Evolution Mechanism of Mining Stress Field

As the mining depth increases, the self-weight stress of the coal body, the structural stress and the mining stress overlap, forming a very complex stress situation in the mining field, causing a series of changes in the physical and mechanical properties of the rock, and showing obvious strong disturbance characteristics. Some problems related to rock mechanics science and engineering gradually change from quantitative to qualitative. The deformation and destruction degree and mode of coal and rock mass are quite different from those in shallow parts. The existing theories and tunnel support methods applicable to shallow mining have encountered problems in deep engineering practice. The resource extraction is extremely difficult and the risk of major mine safety accidents increases, seriously threatening the safe production of the mine. Therefore, it is necessary to explore the temporal and spatial evolution laws of deep mining fields and the loss evolution mechanical characteristics of coal and rock masses.

3.1 Definition of Deep Coal Mining

How to determine the depth of coal mining, previous studies have given different interpretations from multiple angles. "Deep" is a mechanical state determined by the level of geostress, mining stress state and surrounding rock properties. The deep coal mining environment can be summarized as "three highs" (high geostress, high geotemperature, high karst water pressure) and deep rock engineering response "three strong" (strong rheology, strong wet heat, strong dynamic disaster) characteristics [1, 2]; at this time, the deep engineering rock mass has nonlinear mechanical characteristics, and the existing linear mechanical system theory and technology are partially or completely invalid. Deep engineering surrounding rock control can be based on rock mechanics characteristics and engineering characteristics, using difficulty coefficient and danger index as evaluation indicators of stability difficulty [3]. At present, the definition methods of deep mining are from the depth of geostress field characteristics, absolute mining depth, mining coal and rock geostress environment, the degree and mode of disaster caused by mining, tunnel support and maintenance cost and rock mechanics state, etc. The characteristics of deep mining have been diagnosed from the outside to the inside, and targeted solutions for deep mining theory and practice have been proposed. Scientifically defining the depth is an important issue in the development of deep mining theory and practice. It is of great significance to explore the definition method of deep mining suitable for modern coal mining practice in China. For this reason, Professor Zhang [4] combined the deep rock and groundwater environment of coal mine areas in China and modern mining methods, combined the analysis of regional stress field and mining stress field, based on the analysis of shallow crust and quasi-static water stress state in deep coal mine areas in China, further studied the definition of deep in coal mine areas in China, the relative deep definition based on different coal and rock states (rock properties and combinations, water content, etc.) and the method of determining dynamic deep area during mining.

3.1.1 Deep Coal Mining

The manifestation of deep mechanical state is the basic condition for coal mining to enter the deep from the shallow, and the high geostress environment and the nonlinear mechanical response of the original rock are the basic characteristics of the deep mechanical state. Therefore, compared with shallow coal mining, deep coal mining refers to mining activities in the coal and rock space with high geostress environment and mining nonlinear mechanical response. Its connotation mainly includes:

1. Deep mining refers to mining activities where the bedrock is under high ground stress at depth. High ground stress is a basic characteristic of deep stress. At present, the average mining depth of the main mines in the east has reached 800–1000 m, while the western mining area has gradually entered 400–700 m from 100–300 m. Compared with shallow mining, mining in different regions gradually enters a high ground stress environment as it shifts to greater depths. At this time, the stress state of the bedrock gradually shifts from being dominated by tectonic stress to being dominated by vertical stress, and enters a deep stress state when it enters a triaxial compressive stress state with equal pressure in two directions (or quasi-hydrostatic stress state) [5].
2. Deep mining is a mining activity where the mined coal rock shows significant non-linear mechanical response characteristics. The non-linear response of mined coal rock is the dynamic characteristic difference between deep and shallow mechanical states. Under deep conditions, the mechanical response of coal rock transitions from purely elastic deformation to brittle-plastic deformation and plastic flow state, with large plastic deformation, dynamic disasters, and large-scale dynamic instability of the surrounding rock appearing during mining.

3.1 Definition of Deep Coal Mining

Compared with shallow mining, elastic deformation, brittle-plastic deformation, and plastic deformation phenomena coexist, and traditional linear theories and methods are difficult to explain.

3. The process of deep mining is also a process of mining-induced coupling and the evolution of the mechanical state of coal rock. The initial state of the mined coal rock reflects the static properties and mechanical state of the mined coal rock, the mining-induced coupling state reflects the dynamic properties and mechanical state of the coal rock, and the change in the mechanical state of the mined coal rock is related to depth, the combination of bedrock lithology and mining source parameters. Compared with shallow mining, not only is the initial state of the mined coal rock in a quasi-hydrostatic pressure environment considered deep mining, but also the space that appears in the mining process with deep mechanical state is considered a deep area, and the dynamic high-stress area and coal rock non-linear mechanical response that appear at this time also need to be explained by deep mining theory and methods.

Therefore, according to the distribution of coal mine areas in China and the characteristics of coal rock mass lithology combination [6–9], deep mining can be divided into deep mining in the middle and eastern parts and deep mining in the west. The former mainly has the coal-forming period of Carboniferous-Permian, and the coal-bearing strata are mainly Permian Shanxi Formation and Carboniferous Taiyuan Formation, etc.; the latter mainly has the coal-forming period of Jurassic, and the coal-bearing strata are mainly Jurassic Yan'an Formation.

3.1.2 Deep Judgment Criteria and Determination Methods

1. Judgment criteria

The deep mechanical state serves as the main indicator for judging the entry into deep mining, and determining the characteristic depth based on experimental measurement or deep rock in-situ testing is an important breakthrough in the theoretical study of deep mining. The deep state is closely related to the regional ground stress level, mining geological environment, and mining dynamic behavior, and the randomness and dispersion of the sampling of the mine area's ground stress research make the definition of the deep boundary in the mine area limited. It is reasonable to determine the regional depth based on the trend of stress field changes in China's shallow crust and coal mine areas, and to determine the mining area's depth in combination with the characteristics of the rock lithology combination in the mining area and the change rule of the mining-induced coupling mechanics time–space response, thereby defining whether mining has entered the deep.

The study of the regional stress field in the shallow part of China's crust is mainly based on the ground stress testing data of sedimentary rocks, igneous rocks, and metamorphic rocks. The original rock lateral pressure coefficient K_H, K_h, and K_{aV} (that is, the ratio of the maximum horizontal principal stress, the minimum horizontal

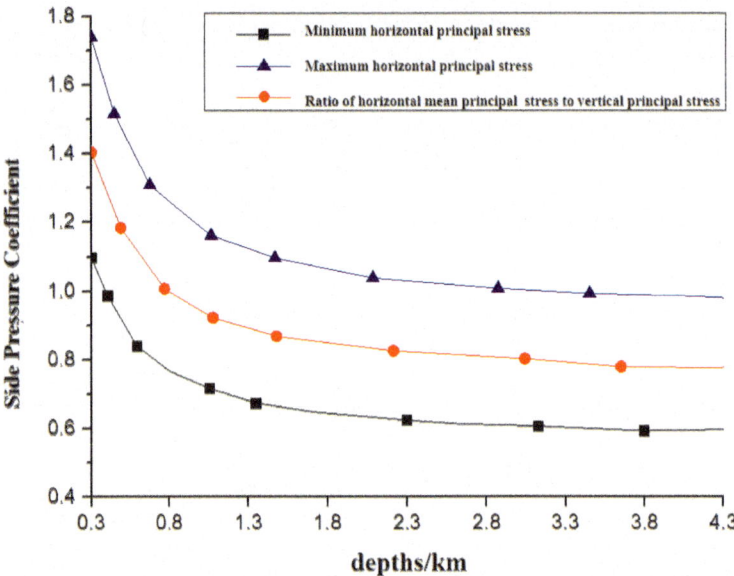

Fig. 3.1 Trend of lateral pressure coefficient changes in shallow crustal regions of China [4]

principal stress, and the average horizontal principal stress to the vertical principal stress) and the statistical distribution law with depth [10] show (Fig. 3.1), shallow K_H and K_h The range of variation is large, which means that local construction stress is dominant. As the depth increases, the measured values converge relatively.

Since K_{aV} integrates the triaxial stress parameters of the original rock, reflecting the relationship between the average horizontal stress and vertical stress, $K_{aV} \approx 1$ approximately reflects the quasi-hydrostatic stress state of the original rock triaxial stress relationship. Therefore, within the currently measurable and minable depth range, using the deep quasi-hydrostatic pressure environment and K_{aV} parameters as the criteria for judging whether coal mining has entered the deep part is reasonable and applicable. At the same time, referring to the average geostress level of coal mines in China and studying the local geostress level differences of coal-bearing strata in the mining area, it is necessary to distinguish and define the specific range of the deep part.

2. Determination method

The deep mechanical state is divided into static and dynamic states. The former is the comprehensive manifestation of the basic physical properties under the deep original rock state, and the latter is the dynamic manifestation when the original rock and the mining source are coupled. According to the deep judgment criteria, whether the mining has entered the deep part and the critical depth can be determined based on whether K reaches the quasi-hydrostatic stress state during the static and dynamic mechanical states.

3.1 Definition of Deep Coal Mining

(1) Determination of static (without mining disturbance) depth

The static depth refers to the area V_m^0 that exhibits a deep mechanical state when the mining system is static ($t = 0$), H_m is the critical depth of the deep part, and the original rock in this area is called deep original rock. Its deep mechanical state function is:

$$K_{aV}^0 = f_S(x, y, z, \sigma, F)|_{Z>H_m}$$

In the deep area V_m the original rock stress field is dominated by horizontal construction stress, showing a shallow state; the original rock stress field in the deep area V_m shows a deep quasi-hydrostatic stress state.

When defining the static deep area, the original rock parameter F is the main influencing factor, that is, the composition and physical properties of the original rock, the lithology combination and the water content of the rock layer have an important impact on the definition of the deep state. Based on the statistical law of the stress field in coal mining areas in China, combined with the coal mining practice in the eastern part, coal mining areas in China choose the depth of 850–900 m near $K_{aV} \approx 1$ and the relative change with depth is less than $10^{-4}/\text{m}$ as the reference deep critical depth H_m. The actual deep critical depth H_S (or visual deep critical depth) that meets the deep state in the specific mining area is related to the difference in the original rock and regional rock combination and physical properties of the mining area.

(2) Determination of dynamic (during mining disturbance) depth

The dynamic depth refers to the spatial area V_m where the original rock exhibits a deep mechanical state (the K_{aV} approaches 1.0) when the mining system is dynamic ($t > 0$), and the deep mechanical state function can be expressed as:

$$K_{aV}^1 = f_S(x, y, z, \sigma, F, C)|_{Z>H_m}^{K_{aV} \approx 1}$$

The deep area V_m includes the static deep area V_m^0 and the dynamic deep area V_d. Outside V_m the lateral pressure parameter $K_{aV} > 1$, the stress state of the original rock and mining coupling is shown as a shallow state, and the stress field is dominated by horizontal construction stress; inside V_m, the original rock physical parameter F and the mining source parameter C jointly determine the stress field state and deep range V_d.

If K_{aV}^R is set as the original rock stress state controlled by the regional (referring to the coal mining area) stress field, ΔK_{aV}^c is the mining increment generated by the mining coupling effect, then the mining stress field state function (the stress field state function within the mining influence range) ΔK_{aV}^s is:

$$K_{aV}^s = K_{aV}^R + \Delta K_{aV}^c$$

At this time, $\Delta K_{aV}^c > 0$ The area shows the shallow mechanical state, the mining coupling action area has increased structural stress or relative decrease in vertical stress, dominated by structural stress; in $\Delta K_{aV}^c < 0$ and $H_S > H_m$ area $\left(K_{aV}^s(x, y, z) = K_{aV}^R(x, y, H_m)\right)$, the mining coupling action shows a decrease in horizontal structural stress or an increase in vertical stress, and locally there may also be areas that conform to the deep mechanical state.

3. Progress in the study of deep mining definition

Academician Kang [11] summarized the geostress measurement data obtained from more than 20 mining areas and 395 measurement points over the years by the Mining Design Research Branch of the Coal Science Research Institute using small-diameter hydraulic fracturing geostress measurement devices. He studied the distribution characteristics of deep and shallow geostress, and believes that: in China, the structural stress of shallow rock layers (burial depth less than 250 m) is significantly dominant, and the geostress state is $\sigma_H > \sigma_h > \sigma_V$, σ_H/σ_V values are between 1.5–2.5; the geostress state of medium-depth mining areas (burial depth is 250–600 m) is generally $\sigma_H > \sigma_V > \sigma_h$, σ_H/σ_V values are between 1.0–2.0; when the burial depth exceeds 600 m, the vertical stress generally dominates, and the geostress state is $\sigma_V > \sigma_H > \sigma_h$, σ_H/σ_V values are between 0.5–1.5; however, in mining areas significantly affected by geological structures, the geostress may still be dominated by structural stress, in a $\sigma_H > \sigma_V > \sigma_h$ state.

When mining work enters the deep part, the environment in which the rock is located has changed significantly, and its mechanical properties have also changed greatly. As the burial depth of the rock increases, the original rock stress that the rock is subjected to also increases, even exceeding the uniaxial compressive strength of the rock itself; at the same time, the deep rock mass contains traces of tectonic movement and contains a higher stress field; in addition, the stress concentration caused by roadway excavation and coal seam mining will also increase the internal stress of the surrounding rock, which may even be much greater than the original rock stress. Compared with the mechanical properties of shallow rock masses, the mechanical properties of deep rock masses have changed greatly [12–14]. The main difference between deep mining and shallow mining lies in the special environment in which deep rocks are located, that is, the complex mechanical environment of "three highs and one disturbance". The high-stress environment in the deep part is the decisive factor for engineering disasters in deep resource mining.

The deformation mechanics theory (continuum theory) established under shallow mining conditions all follow a basic assumption, that is, the object is continuous, which means that the volume of the entire object is continuously occupied by the material elements that make up this object. Under this premise, some physical quantities of object movement, such as stress, deformation, displacement, etc., may be continuously changing, and they can be represented by continuous functions of coordinates to express their changing laws. However, in deep engineering, the limit situation defined above does not actually exist. Because the rock has discontinuity at the grain size range, and tends to a certain value smaller than the grain or molecular spacing. The theoretical stress is the stress at a certain point, and the stress in the material world is the average stress of a certain element, and the rock mass itself contains many micro-cracks, cavities, joints, and the material organization has non-uniform and discontinuity. This shows that in the field of deep rock mechanics, we cannot directly borrow the continuity assumption and definition of classical theoretical mechanics, and use the theory of continuum mechanics to analyze the deep rock mechanics problems of highly discontinuous media, and we must consider the reasonable use range of assumptions and the applicable definition of various physical quantities.

Academician Xie [15] It is pointed out that due to the inherent attributes of the typical "three high" environment of deep rock mass and the additional attributes of "strong disturbance" and "strong timeliness" of resource extraction, frequent occurrences of high-energy, large-volume engineering disasters in deep parts are difficult to predict and effectively control, and the mechanism is unclear. The applicability of traditional rock mechanics and mining theory in deep parts is controversial. The fundamental reason is that after entering the deep part, the nonlinear behavior of rock mass materials is more prominent, and the in-situ stress state of rock mass and the effect of geostress environment are more prominent. The existing rock mechanics theories are all based on the material mechanics based on static research perspective, which is lagging behind human rock engineering practice activities and is irrelevant to depth, engineering activities and deep in-situ environment.

Professor Zhang [4] Through the actual deep critical depth calculation and comparison with the reference deep critical depth, it is shown that the eastern and central mining areas are close to this depth, and the western part is smaller. The comparative study of the actual deep mining critical depth in the typical mining areas in the east, middle and west of our country shows that: compared with the reference deep mining critical depth obtained by the coal mining areas in our country, the eastern mining areas are deeper, the middle mining areas are similar in depth, and the western (Shaanxi, Mongolia, etc.) underground water-rich mining areas are shallower, reaching the actual deep mining critical depth at 500–600 m, and the large mining height working face at both ends of the mining depth of 400–500 m also shows the deep mechanical state.

3.2 Research on Coal Body Damage Evolution Characteristics

3.2.1 Analysis of Coal Rock Damage Mechanism

Coal rock mass is a heterogeneous body composed of various sizes of mineral particles and bonded together by certain binding materials, especially coal body is a porous medium, which contains various sizes of original pores, cracks and other defects and foreign inclusions. The essence of deformation and failure of coal rock mass under external load is the continuous initiation, expansion and new crack generation, evolution and penetration of internal original cracks, until the formation of macro cracks leads to material instability and failure [16–18]. Pores, cracks and other defects can reduce the mechanical bearing capacity of coal rock mass, which can be vividly called damage. Therefore, introducing the concept of damage and the analysis method of damage mechanics into the research of the nucleation, evolution and deterioration process of small pores and cracks in coal rock materials has good adaptability, and the damage constitutive model considering the size, number and distribution of cracks can well explain and guide some practical engineering.

1. Coal Rock Damage Mechanics Theory

As an important branch of solid mechanics, the theory of damage mechanics has a history of more than 100 years since its proposal [19, 20] In 1895, a Soviet mechanic proposed the concept of "damage" while studying the mechanical properties of metal materials. This was the first recorded appearance of the term "material damage" in literature. After defining and explaining material damage, terms such as "continuous damage factor" and "effective stress" began to be introduced into papers on the study of metal material creep fracture to describe the damage state of metal materials. However, the description of metal material damage was still qualitative. Subsequently, based on the concept of damage, the concept of "damage factor" was proposed, and combined with the phenomenological method in continuum mechanics, it was applied to the study of metal material creep damage. In order to combine with mathematical analysis methods, the "damage variable" that can be quantitatively described was proposed from a quantitative perspective to study the damage and destruction of metal materials more deeply, and a relatively systematic explanation and application of this concept was given.

Research on metal material damage is relatively mature, but considering the differences between coal rock materials and metal materials, more factors are considered when defining and studying coal rock material damage. In view of the advantages of the concept of damage in the study of the mechanical properties of metal materials, in 1976, the concept of damage mechanics was combined with the characteristics of rocks, quasi-rocks and concrete materials, and the concept of describing rock damage was explained. Based on the definition of damage and the research methods of damage mechanics, a coal rock damage constitutive relationship model based on continuum theory was established. Chinese scholars have done a lot of research on coal rock damage mechanics, which has attracted wide attention from foreign scholars. Among them, Academician Xie He'ping combined the macroscopic fractal theory with the microscopic damage mechanics, and for the first time proposed the fractal damage mechanics theory, which complements the research of other scholars and together forms a relatively complete theoretical system of coal rock damage mechanics. The development of coal rock damage mechanics has been continuously valued by rock mechanics researchers. At present, the related research content in the study of coal rock damage mechanics can be summarized as shown in Fig. 3.2.

3.2 Research on Coal Body Damage Evolution Characteristics

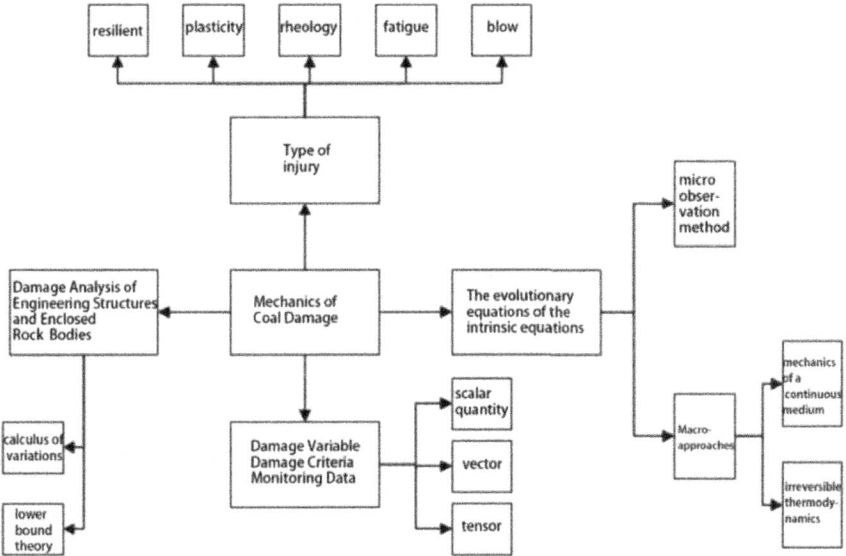

Fig. 3.2 Main research content of damage mechanics

2. Acoustic emission and energy theory of coal rock damage deterioration
(1) Acoustic emission theory

During the damage deterioration process of coal rock materials under external force, the generation of internal cracks releases energy in the form of elastic strain energy, which is called the acoustic emission phenomenon [21, 22]. The acoustic emission law of coal rock reflects the internal damage evolution process of the material. Therefore, the acoustic emission signals released from the material (such as acoustic emission energy, ringing count, impact rate, waveform, damage crack position, etc.) can be collected, recorded and processed by acoustic emission information monitoring equipment, and then the damage evolution process of the material can be inverted and analyzed according to the change law of various parameters of acoustic emission [23–32], thereby further providing a reference for the stability prediction and control of rock engineering such as mining engineering and tunnel engineering.

In terms of studying the damage deterioration characteristics of coal rock materials through acoustic emission parameters, the damage factor definition and derivation of acoustic emission parameters can be carried out. According to the definition of the damage factor, the damage factor of coal rock material D is:

$$D = \frac{V_n}{V} \tag{3.1}$$

where, V_n is the volume of damage such as micro-cracks formed by the damage of coal rock materials; V is the total bearing volume of the material before damage.

When $V_n = 0$, $D = 0$, it indicates that the coal rock specimen is in a state without any damage; when $V_n = V$, $D = 1$, it indicates that the coal rock specimen is in a completely damaged state that cannot bear any external load. Therefore, D is between 0–1, and represents the damage state variable of the coal rock specimen at a certain moment during the external load loading process.

The damage variable D actually determines the elastic modulus of the coal rock specimen E, if the elastic modulus of the initial state without any damage of the coal rock material is E_0, then the elastic modulus corresponding to any damage state E is:

$$E = E_0(1 - D) \tag{3.2}$$

Therefore, the elastic strain energy stored under the action of external load U is:

$$U = \frac{E_0(1 - D)\varepsilon^2}{2} \tag{3.3}$$

where, D is the damage variable of the coal rock specimen at a certain moment; ε is the strain value when the coal rock specimen is damaged D.

According to formulas 3.1, 3.2 and the thermodynamic relationship, the differential constitutive relationship of damage is obtained:

$$\left(\frac{dU}{d\varepsilon}\right) D = E(1 - D)\varepsilon = \sigma \tag{3.4}$$

There are many small original defects in coal rock materials. When the stress exceeds the limit strength it can bear, it will cause the expansion of original defects and the birth, expansion, and evolution of new defects, and at the same time release elastic strain energy in the form of elastic waves, thereby generating acoustic emission phenomena.

The Weibull distribution proposed by Jaeger and Cock is usually used to define the distribution of small defects such as cracks in coal rock materials, that is:

$$n(\varepsilon) = k\varepsilon^m \tag{3.5}$$

$$n'^{(\varepsilon)} = km\varepsilon^{m-1} \tag{3.6}$$

where, $n(\varepsilon)$ is the number of defects that can be caused by strain less than or equal to ε; $n'(\varepsilon)$ is the change rate of defects with strain; k and m are constants, representing the fracture activity properties of the material.

When the strain increases by a certain increment $d\varepsilon$, the number of new cracks and other defects that can be excited is:

$$dn = n'^{(\varepsilon)} d\varepsilon \tag{3.7}$$

3.2 Research on Coal Body Damage Evolution Characteristics

Due to the previous damage, in the material with a percentage of the total volume of D, the elastic strain energy has been released, resulting in a reduction in the number of defects actually activated $(1 - D)$ times, and at the same time, it is assumed that the damage of a defect corresponds to an acoustic emission count, then the acoustic emission count can be expressed as:

$$dN = (1 - D)n'^{(\varepsilon)}d\varepsilon \tag{3.8}$$

From Eqs. (3.2) and (3.7), we get:

$$dN = km(1 - D)\varepsilon^{m-1}d\varepsilon \tag{3.9}$$

$$N = r(\varepsilon)\int_{\varepsilon_0}^{\varepsilon} km(1 - D)d\varepsilon \tag{3.10}$$

where, ε_0 is the strain of the initial damage of the coal rock material; $r(\varepsilon)$ is a random factor, randomly taking values between [0, 1].

Therefore, the acoustic emission rate can be expressed as:

$$N'^{(t)} = \frac{dN}{dt} = r(\varepsilon)km(1-D)\varepsilon^{m-1}\frac{d\varepsilon}{dt} \tag{3.11}$$

Equations 3.10 and 3.11 are the theoretical models of acoustic emission of coal rock materials under external load. From the variables of the model, it can be seen that the acoustic emission count and the acoustic emission rate not only depend on the damage factor D, instantaneous strain ε and strain rate ε', but also closely related to the inherent properties of coal rock (material size, homogeneity of the material, etc.).

(2) Energy theory

The defects such as micro-cracks in the coal rock body continuously germinate, expand, penetrate until they evolve into macro-cracks, leading to the instability and destruction of the coal rock body. The deformation and damage process of coal rock under external force can be regarded as an energy (kinetic energy, elastic strain energy, etc.) driven phenomenon of energy transfer and transformation from one state to another [33]. Therefore, discussing the energy mechanism in the deformation and destruction process of coal rock from the perspective of energy dissipation and conversion can not only ignore the uncertainty factors brought by the intermediate change process, but also more truly and accurately reflect the essence and law of coal rock damage and destruction.

Take the uniaxial compression of coal rock as an example. As shown in Fig. 3.3, as the external force loading continues, energy is continuously input, part of which is dissipated as heat energy, this part becomes irreversible dissipation energy, and part of it is stored as elastic energy, this part becomes elastic strain energy. Among

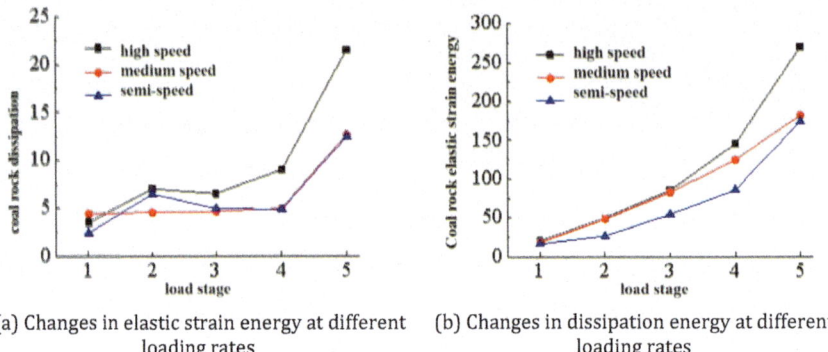

(a) Changes in elastic strain energy at different loading rates

(b) Changes in dissipation energy at different loading rates

Fig. 3.3 Changes in elastic energy and dissipation energy at different loading rates

them, the growth process of elastic strain energy can be divided into three stages (non-linear slow growth → stable linear growth → growth slowdown stage), and the growth of dissipation energy can be divided into two stages (slow growth stage → significant growth stage). Hard rock and soft rock have different proportions of dissipation energy and elastic strain energy in the energy input process under external force. Among them, before the peak failure, most of the energy input into hard rock is converted into elastic strain energy and stored, while the energy input into soft rock is used for damage work and converted into heat energy and other dissipation energy is released. The loading rate has different effects on elastic strain energy and dissipation energy. Among them, as the loading rate increases, the total energy input increases, and the proportion of elastic strain energy also shows an increasing trend, while the dissipation energy shows a trend of first increasing and then decreasing, and the larger the loading rate, the more obvious the evolution process of elastic strain energy with external force loading.

As can be seen from Fig. 3.4, before the yield stress (point c), the sum of friction energy, kinetic energy, strain energy, and bonding energy is basically equal to the boundary energy. The proportion of bonding energy and strain energy is relatively large, this part of the energy is related to the generation and driving of cracks, and its rise and fall is related to the deterioration of the model material. The friction energy representing the action of cracks is the opposite, they are mutually exclusive. Because the generation of cracks has to overcome the bonding energy, and then expand under the drive of strain energy, the friction part only starts to work after the cracks are generated. After the stress reaches the peak strength (point d), the bonding energy and strain energy decrease sharply, the friction energy increases sharply, and the proportion of friction energy gradually increases with the further expansion of the cracks, which shows that the friction effect is the main provider of residual strength. The proportion of kinetic energy in the entire deformation process of the specimen is not large, which is related to the loading process and the internal dynamic balance

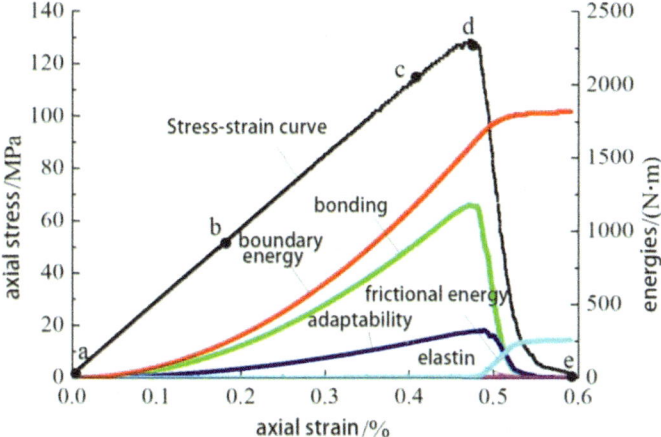

Fig. 3.4 Energy evolution curve of BPM coal rock model

of the specimen, indicating that the deformation of the specimen is not very violent, and the cracks stably expand and penetrate.

3.2.2 Characteristics of Coal Rock Damage Evolution

1. Constitutive model of coal rock damage evolution-based on acoustic emission theory Kachanov [34] defined the damage variable as:

$$D = \frac{A_d}{A} \qquad (3.12)$$

where, A_d for the cross-sectional area of the material at a certain moment of damage; A for the cross-sectional area of the material at the initial moment of no damage.

Assuming the cross-sectional area of the undamaged material A The cumulative number of acoustic emissions at the moment when the load-bearing capacity is completely lost is S, then the cumulative count of acoustic emissions per unit area of coal rock damage S_W is:

$$S_W = \frac{S}{A} \qquad (3.13)$$

When the cross-sectional damage reaches A_d, the cumulative count of acoustic emissions S_d is

$$S_d = S_w A_d = \frac{S}{A} A_d \qquad (3.14)$$

Therefore, we have

$$D = \frac{S_d}{S} \qquad (3.15)$$

The uniaxial compression damage constitutive model of coal rock material based on acoustic emission characteristics and strain equivalence principle is:

$$\sigma = E\varepsilon(1 - D) = E\varepsilon\left(1 - \frac{S_d}{S}\right) \qquad (3.16)$$

Similarly, the damage variables based on boundary energy and friction energy can be defined as D_b and D_f:

$$D_b = \frac{B_d}{B} \qquad (3.17)$$

where, B_d is the cumulative count of boundary energy when the cross-sectional damage reaches A_d; B is the cumulative count of boundary energy at the moment when the undamaged cross-sectional area of coal rock material A completely loses its load-bearing capacity.

$$D_f = \frac{F_d}{F} \qquad (3.18)$$

where, F_d is the cumulative count of friction energy when the cross-sectional damage reaches A_d; F is the cumulative count of friction energy at the moment when the undamaged cross-sectional area of coal rock material A completely loses its load-bearing capacity.

The uniaxial compression damage constitutive models of coal rock material based on boundary energy and friction energy characteristics are obtained as follows:

$$\sigma = E\varepsilon(1 - D_b) = E\varepsilon\left(1 - \frac{B_d}{B}\right) \qquad (3.19)$$

$$\sigma = E\varepsilon(1 - D_f) = E\varepsilon\left(1 - \frac{F_d}{F}\right) \qquad (3.20)$$

Figure 3.5 is the stress–strain curve of coal rock material fitted based on acoustic emission, boundary energy and friction energy parameters. As can be seen from the figure, the constitutive model fitted based on friction energy parameters can best reflect the stress–strain characteristics of coal rock material, followed by the constitutive model based on acoustic emission parameters. The reason for the large

3.2 Research on Coal Body Damage Evolution Characteristics

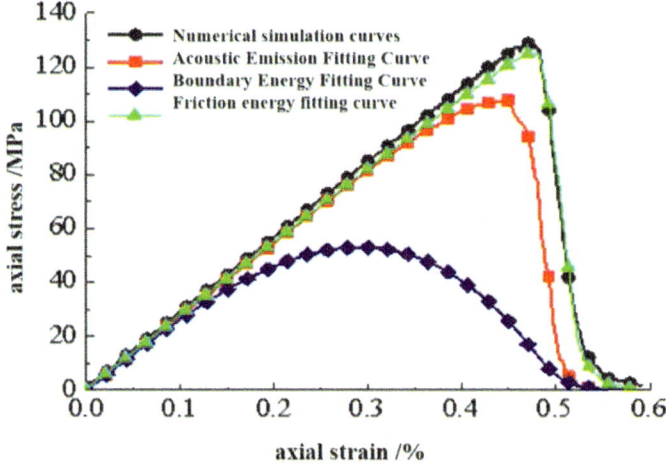

Fig. 3.5 Fitting coal rock stress–strain curve

difference between the constitutive model of coal rock material based on boundary energy characteristics and the numerical curve is that the boundary energy needs to provide a large amount of energy for model damage at the beginning, and the model absorbs it through bond energy and denaturation energy, without causing damage. Since it is difficult to record the change characteristics of friction energy during the compression process of actual coal rock material specimens, the characteristics of acoustic emission can be used to study the damage characteristics of coal rock material.

2. Coal rock damage evolution constitutive model-based on strain energy theory [35]

According to the second law of thermodynamics, the failure of rock is caused by the energy applied to the material exceeding the limit value that the material can withstand. Rock material is a heterogeneous and discontinuous material, and the mechanical properties of rock failure show complex changes under different geostresses and engineering forces. Local high stress and high strain cause local damage to the rock, loss of strength, but not necessarily lead to the overall failure of the rock mass. The existing classical plasticity theory-based rock failure criteria are difficult to analyze the strength changes and overall failure behavior under complex stress conditions. Therefore, analyzing the deformation and failure properties of rock from the perspective of energy is more conducive to reflecting the essential characteristics of rock strength changes and overall failure [36–38]. Based on the assumption that the elastic modulus of rock microscopic units approximately follows the Weibull distribution, combined with the theory of strain energy density, a rock damage constitutive model was established. The homogeneity coefficient of rock m and the elastic modulus reduction coefficient K_0 were determined using the energy signal of acoustic emission events and the longitudinal wave velocity of rock. The established damage constitutive model was used for uniaxial loading simulation. The simulation result curve was compared with the theoretical curve of the existing model and the experimental curve of uniaxial loading. The model can well describe the stress–strain relationship and acoustic emission situation of the specimen. The construction of the rock damage constitutive model based on strain energy density provides a new theoretical basis for comprehensively considering the homogeneity of rock and the impact of repeated loading processes on rock specimens.

1) Damage constitutive model based on Weibull distribution

There are a large number of pores and cracks in the rock mass, which divide the rock mass into micro units. The mechanical properties of each unit are not equal and are randomly distributed. Based on the elastic modulus of rock units following the Weibull distribution, a rock damage constitutive model was established. This model can reflect the stress and strain during the loading process, as well as the acoustic emission characteristics. The specific model is

$$\sigma_i = \sigma_i'(1 - D) = E\varepsilon_i(1 - D) \tag{3.21}$$

where, σ_i' is the actual stress, σ_i is the nominal stress, D is the rock damage variable, ε_i is the strain, E is the elastic modulus.

The elastic modulus of the unit degree approximately follows the Weibull distribution, and the damage variable D can be determined by the statistical damage theory, and its probability density is

$$P(x) = \frac{m}{F_0 \left(\frac{F}{F_0}\right)^{m-1} e^{-\left(\frac{F}{F_0}\right)^m}} \tag{3.22}$$

where, m is the homogeneity coefficient, F is a function of the elastic energy density of the unit body, F_0 is the value corresponding to F when completely damaged.

$$F = f(W) = \int_0^{\varepsilon_{ij}} \sigma_{ij} d\varepsilon_{ij} \tag{3.23}$$

The rock damage parameter D can be expressed as

$$D = \int_0^F P(x)dx = 1 - \exp\left[-\left(\frac{F}{F_0}\right)^m\right] \tag{3.24}$$

When the rock deforms under the action of external forces, the unit will store strain energy. Each unit can store a limited amount of strain energy. After exceeding the limit of strain energy that the unit can withstand, the unit will be damaged and the elastic modulus will also decrease. To describe the change of the elastic modulus

3.2 Research on Coal Body Damage Evolution Characteristics

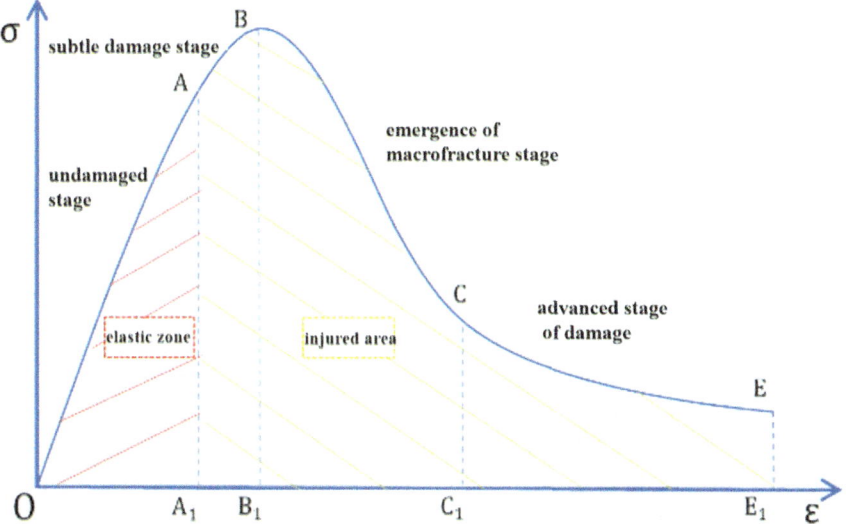

Fig. 3.6 Rock damage zoning diagram

under the action of external forces, the loading process is divided into zones according to the strain energy stored in the unit body (Fig. 3.6).

When the strain energy density of the unit is less than S_{OAA_1}, the unit is in the elastic stage and no damage has occurred.

When the strain energy density of the unit is greater than S_{OAA_1}, the unit is damaged, the elastic modulus of the rock decreases. To describe the change of the elastic modulus in this stage, Sun and others [39] discretized the reduction of the elastic modulus, setting the maximum number of reductions to $n = 20$ times, and each reduction coefficient remains unchanged at K_0 (Fig. 3.7). $E_n = (K_0)^n E$, so the unit body elastic density function:

$$F = \int_0^{\sigma_{ij}} \varepsilon_{ij} d\sigma_{ij} \approx \frac{1}{2} \sum_{n=1}^{20} \frac{\sigma_{ij}^2}{E_n} \tag{3.25}$$

2) Parameter correction of the damage constitutive model

In order to determine the parameters in the damage constitutive model, Li and others [40] The extreme values of the stress–strain curve under unconfined pressure are used to determine the relevant parameters in the damage constitutive model. This method can fit the stress–strain relationship under different confining pressures well. However, the parameters selected by this method are related to the confining pressure and it is difficult to reflect the stress–strain situation under complex stress conditions such as repeated loading. Based on the energy density theory, numerical simulations are used to conduct numerical experiments on the damage process of the rock

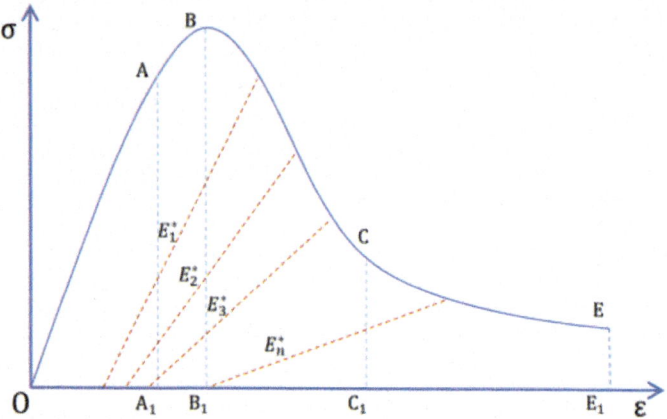

Fig. 3.7 Schematic diagram of elastic modulus reduction principle

mass, and the simulation results under different parameters are compared with the laboratory experimental results to determine the relevant parameters.

(1) Numerical simulation method of strain energy density theory

Using the FLAC3D finite difference numerical software, the damage equation of the rock unit is established according to the above theory, and the simulated specimen is a standard cylindrical specimen (diameter D = 50 mm, height L = 100 mm). The volume modulus and shear modulus of the unit follow the Weibull distribution. The model uses velocity control to load the top and bottom. In FLAC3D, the external parameter Zextra1 of the unit is the number of unit failures, Zextra2 is the strain energy of the unit, and Zextra3 is the failure energy threshold of the unit. Using the built-in FISH language, the energy of each unit is set to 0 before the calculation starts, that is, Zextra2 = 0, and the strain energy density of the unit at the first step is 0. When calculating to the i-th step, the strain energy density of the unit is:

$$(dW/dv)_i = (dW/dv)_{(i-1)} + \frac{1}{2}(\sigma_x^i + \sigma_x^{i-1})(\varepsilon_x^i - \varepsilon_x^{i-1})$$
$$+ \frac{1}{2}(\sigma_y^i + \sigma_y^{i-1})(\varepsilon_y^i - \varepsilon_y^{i-1}) + \frac{1}{2}(\sigma_z^i + \sigma_z^{i-1})(\varepsilon_z^i - \varepsilon_z^{i-1})$$
$$+ \frac{1}{2}(\sigma_{xy}^i + \sigma_{xy}^{i-1})(\varepsilon_{xy}^i - \varepsilon_{xy}^{i-1}) + \frac{1}{2}(\sigma_{yz}^i + \sigma_{yz}^{i-1})(\varepsilon_{yz}^i - \varepsilon_{yz}^{i-1})$$
$$+ \frac{1}{2}(\sigma_{yz}^i + \sigma_{yz}^{i-1})(\varepsilon_{yz}^i - \varepsilon_{yz}^{i-1}) \tag{3.26}$$

When the strain energy density of the unit body is greater than the failure energy threshold of the unit, it is considered that an acoustic emission event has occurred, and the value of the external parameter Zextra1 of the unit increases by 1, and the elastic modulus of the unit is reduced. If the external parameter Zextra1 of the unit

3.2 Research on Coal Body Damage Evolution Characteristics

body is 20, it is considered that the unit has been completely destroyed, and the elastic modulus of the unit is no longer reduced.

The initial damage threshold of the unit follows the Weibull distribution. The elastic energy of 0.034 MPa is taken as the average initial damage threshold. The elastic modulus decreases continuously with damage, so the threshold of the unit will increase after each damage and become 1/K0 times the previous one. During unloading, the elastic energy of the unit will decrease continuously with unloading, and the elastic energy damage threshold remains unchanged, thereby simulating the Kaiser effect.

In order to prevent the error brought by the grid division to the simulation results, we divide the standard specimen with a diameter D = 50 mm and a height L = 100 mm into 1600 units, 64,000 units, 80,000 units, 100,000 units, and 75,000 units. Among them, when the unit is between 64,000 and 100,000, the difference between stress–strain curve and acoustic emission characteristics is small (as shown in Figs. 3.8 and 3.9) and the influence of grid division on simulation results can be ignored, so 80,000 units are used to simulate the specimen. The cumulative energy of acoustic emission when the specimen is completely damaged is taken as the normalized value. The numerical simulation parameters are shown in Table 3.1.

Since the elastic modulus of the unit is different, different combinations of the relative positions of the units will cause changes in the uniaxial compressive strength of the specimen. In order to study the impact of different combinations of the relative positions of the units on the overall mechanical properties of the specimen. Different random seed numbers (set random) are set in the simulation software to simulate the impact of the relative position of the unit on the overall mechanical properties of the specimen. Table 3.2 shows four situations of the relative position of the unit.

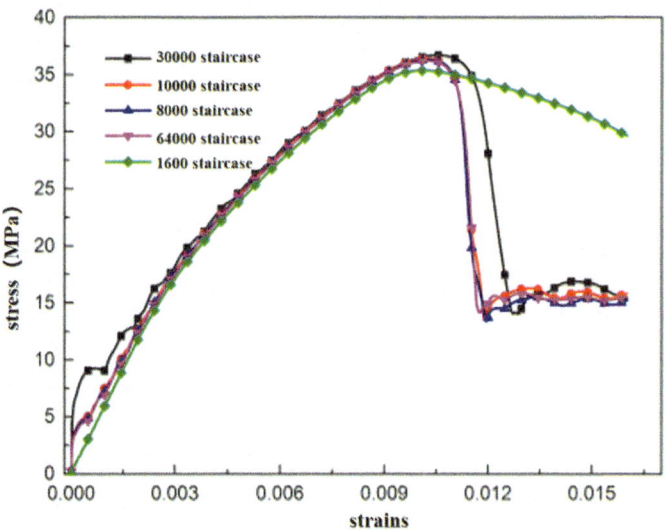

Fig. 3.8 Stress–strain curve under different grid divisions

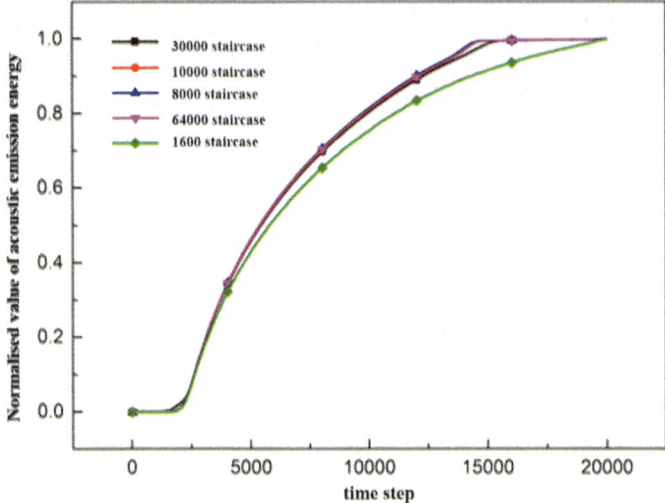

Fig. 3.9 Cumulative acoustic emission signal under different grid divisions

Table 3.1 Mechanical parameters of numerical simulation

Bulk modulus (GPa)	Shear modulus (GPa)	Internal friction angle (°)	Tensile strength (MPa)	Cohesion (MPa)
3.18	2.8	41	8.67	5.82
1600 zones	64,000 zones	80,000 zones	100,000 zones	300,000 zones

The relative position of different units has a small impact on the uniaxial compressive strength and strain of the specimen, only having a certain impact on the post-peak curve (Fig. 3.10). Therefore, the relative position of the units in combination mode 1 is used to simulate the gypsum specimen.

(2) Homogeneity Determination

The heterogeneity of rock has a significant impact on its properties. In the Weibull distribution, m represents the homogeneity of the function. The larger the m, the higher the homogeneity, and the closer the mechanical properties of each unit. To consider the impact of the concentration of unit distribution on the specimen, change the value of m, respectively take m = 3, m = 5, m = 7, m = 9, m = 11, m = 13, m = 15, m = 17, m = 19 to simulate the uniaxial compression of rock, and compare the

3.2 Research on Coal Body Damage Evolution Characteristics

Table 3.2 Unit combination mode

Illustration	Combination mode 1 (set random = 1)	Combination mode 2 (set random = 2)	Combination mode 3 (set random = 5)	Combination mode 4 (set random = 10)

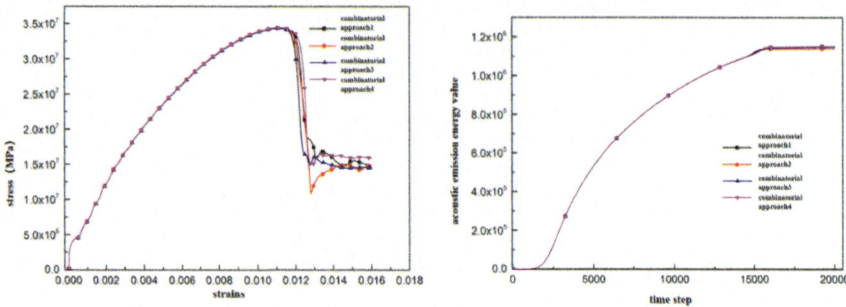

Fig. 3.10 Stress–strain curves and acoustic emission event curves of different unit relative positions

stress–strain characteristics and acoustic emission events. Under the same average parameters of the unit, the higher the homogeneity of the rock, the higher the uniaxial compressive strength of the rock, the smaller the peak strain, the larger the elastic modulus, and the specimen as a whole shows stronger brittleness (Fig. 3.11). To study the impact of homogeneity m on uniaxial compressive strength, fit the curve (Fig. 3.12), the fitting relationship function is:

$$y_1 = 3.6865 \times 10^7 - 15.0459 \times 10^6 e^{\frac{-m}{3.739}} \tag{3.27}$$

In the formula, m is the homogeneity, y_1 is the uniaxial compressive strength.

Under axial compression, rocks with low homogeneity require a large difference in energy for unit destruction. The units are gradually destroyed during the loading

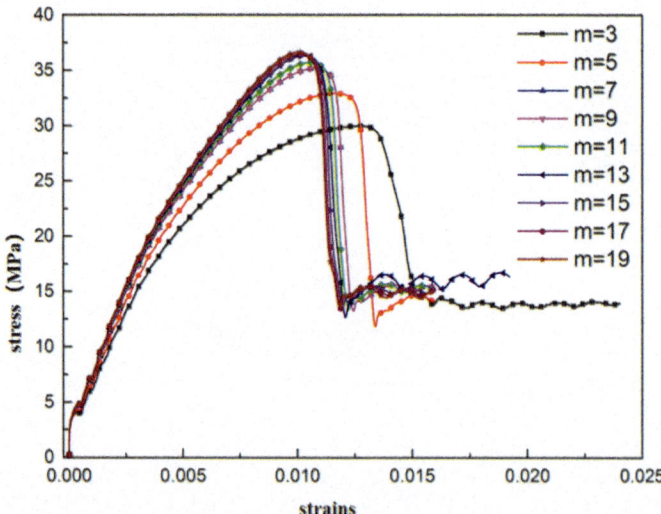

Fig. 3.11 Stress–strain curves under different homogeneities

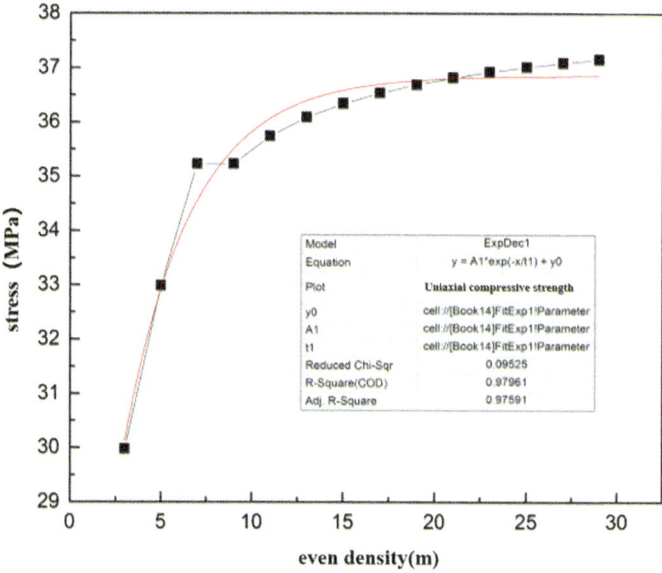

Fig. 3.12 Uniaxial compressive strength under different homogeneities

process, and the specimen absorbs more elastic energy. The more acoustic emission events, the more energy consumed during damage (Figs. 3.13 and 3.14). The lower the homogeneity, the earlier the acoustic emission events occur, the longer the duration, and acoustic emission events occur throughout the compression process of the specimen; while the acoustic emission events of specimens with higher homogeneity appear more concentrated, and there are almost no post-peak acoustic emission events. To study the impact of homogeneity m on acoustic emission events, fit the curve (Fig. 3.14), the fitting relationship function is:

$$y_2 = 1.067 \times 10^6 - 8.5487 \times \frac{10^5}{\left[1 + \left(\frac{m}{0.66309}\right)^{1.02377}\right]} \quad (3.28)$$

In the formula, m is the homogeneity, y_2 is the total number of acoustic emission events.

To determine the appropriate homogeneity m, relevant scholars [41–44] Adopting a nonlinear elastic constitutive relationship, the stress–strain curve relationship is as follows:

$$\sigma = E_0\varepsilon(1 - D) = E_0\varepsilon\exp\left(-\left(\frac{\varepsilon}{\varepsilon_0}\right)^m\right) \quad (3.29)$$

where, ε is the infinitesimal strain, ε_0 is the statistical average strain of the rock material.

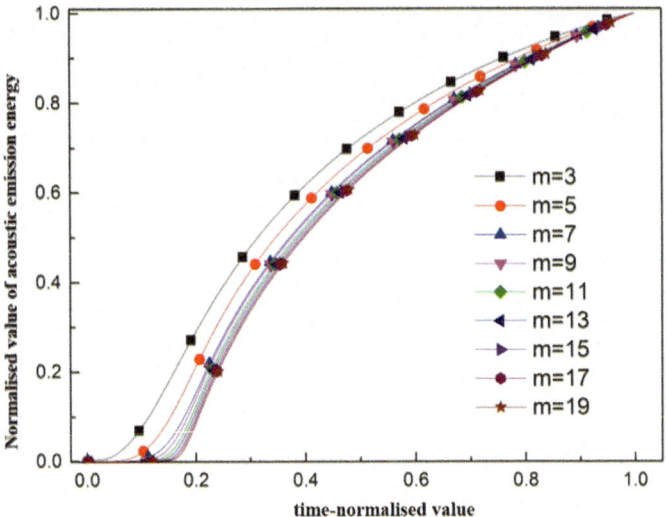

Fig. 3.13 Acoustic emission event-time curves under different homogeneities

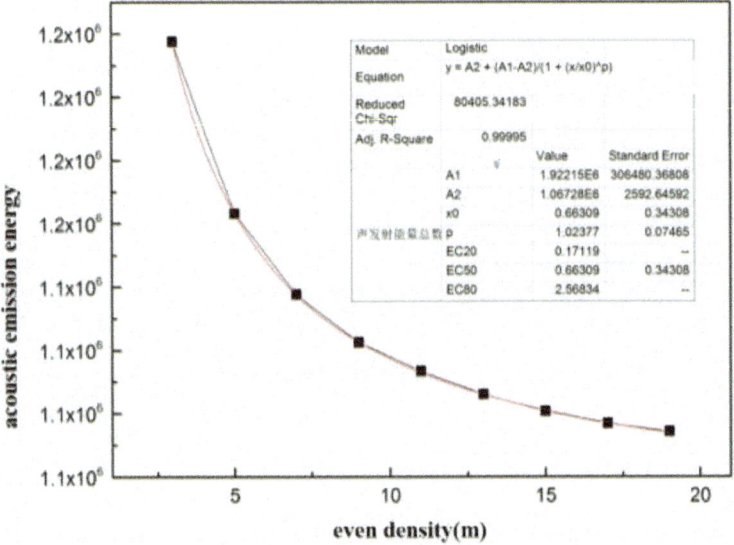

Fig. 3.14 Total number of acoustic emission events under different homogeneities

The calculation results are:

$$m = \frac{1}{\ln\left(\frac{E_0 \varepsilon_c}{\sigma_c}\right)} \tag{3.30}$$

3.2 Research on Coal Body Damage Evolution Characteristics

where, E_0 is the initial modulus of the rock, σ_c is the peak strength, ε_c is the strain corresponding to the peak strength. According to the selected gypsum specimen uniaxial compression results, σ_c is 35 MPa, ε_c is 0.009, E_0 is 4.595 GPa, the calculated $m \approx 5.99$. Zhang [45] considered the post-peak energy consumption and release situation and modified the formula:

$$m = \frac{\ln\left(-\ln\frac{\sigma_c}{E_0\varepsilon_c}\right)}{\ln\left(\frac{\varepsilon_c}{\varepsilon_1}\right)} \tag{3.31}$$

where, ε_1 is the strain when the rock undergoes unstable failure, 0.0126.

The calculated $m \approx 5.32$. It can be seen that considering the post-peak strength curve of the rock, the homogeneity of the rock will correspondingly decrease. Previous studies on homogeneity were mainly based on the characteristics of the stress–strain curve, but strain and damage are not linearly corresponding. Using strain directly as the rock damage index lacks certain rationality.

To determine the homogeneity of the specimen m, through experimental observation, the AE energy is consistent with the statistical distribution of internal damage in the specimen. The damage caused under external load is represented by AE energy. The cumulative AE energy is denoted as C, and the cumulative AE energy at complete damage is set as C_m. The ratio of the two is used to represent the damage situation: $D = C/C_m$, hence $\frac{C}{C_m} = 1 - \exp\left[-\left(\frac{F}{F_0}\right)^m\right]$. Therefore, we can analyze the degree of damage to the specimen by measuring the acoustic emission energy.

The rock taken from the No. 8 mine of Dahanshi Gypsum Mine in Lanling County, Shandong Province was cored and polished to prepare 6 standard specimens with a diameter of D = 50 mm and a height of L = 100 mm. They were numbered S-1, S-2, S-3, S-4, S-5, S-6. The non-parallelism and non-verticality of the specimen ends are both less than 0.02. The S-1 specimen was subjected to a uniaxial compression test, and real-time monitoring was carried out using the PAC16 channel acoustic emission device.

Combining the influence of homogeneity on uniaxial compressive strength and acoustic emission event energy, the difference between the normalized acoustic emission simulation value and the acoustic emission event energy curve obtained from the experiment under different homogeneity is compared (Fig. 3.15), and the square error is calculated (Table 3.3). The smaller the square error, the better the fit.

Under the same mechanical parameter conditions, when $m = 7$, the simulation results and experimental results are the best fit, and also better than the fit degree when $m = 5.32$. The elastic modulus reduction model based on strain energy can more realistically simulate the stress–strain and acoustic emission conditions of the specimen.

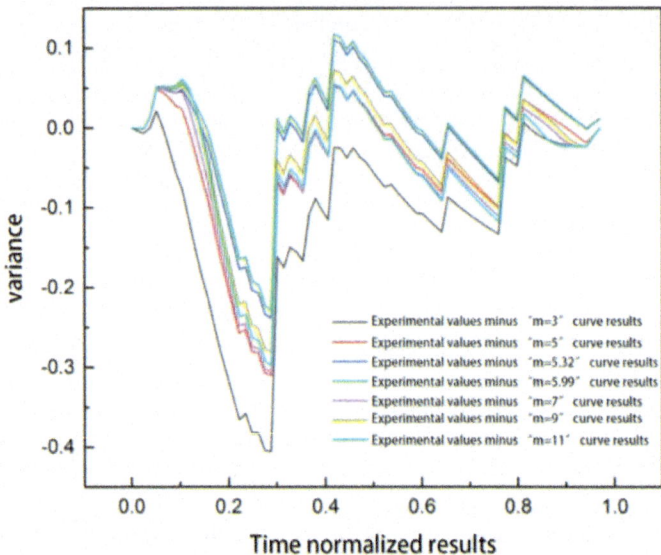

Fig. 3.15 Difference between simulation and experimental values under different homogeneity

Table 3.3 Fitting situation of different homogeneity

Homogeneity (m)	3	5	5.32	5.99	7	9	11	13
Goodness of fit	0.7214	0.4414	0.4214	0.3992	0.3741	0.3742	0.3812	0.3904
Homogeneity (m)	15	17	19	21	23	25	27	29
Goodness of fit	0.4031	0.4145	0.4261	0.4261	0.4261	0.4399	0.4573	0.4634

(3) Determination of the reduction value of the elastic modulus

With the increase of external force, micro-cracks appear in the rock mass, the rock mass unit is damaged, and the elastic modulus of the unit body will decrease accordingly. For the reduction of the elastic modulus, Sun [39] discretized the reduction of the elastic modulus, setting the maximum number of reductions to $n = 20$ times, and the reduction coefficient remained unchanged each time K_0. This method is based on the fitting value obtained from laboratory experiments and has good laboratory application value, but it does not give the physical meaning of the decrease in the elastic modulus. In order to determine the reduction value of the elastic modulus, the relationship between the longitudinal wave velocity of the rock mass and the dynamic bulk modulus and dynamic shear modulus is used:

$$V_P = \sqrt{\frac{K + 1.33G}{\rho}} \tag{3.32}$$

3.2 Research on Coal Body Damage Evolution Characteristics

where, V_P is the longitudinal wave velocity of the rock, K is the dynamic bulk modulus of the rock, which is 2.16 times the static bulk modulus, G is the dynamic shear modulus of the rock, which is 2.16 times the static shear modulus [46, 47], ρ is the density of the rock.

It is assumed that each unit can be reduced up to 20 times during the damage process, and the elastic modulus reduction coefficient remains unchanged each time, all being K_0, and the unit density remains unchanged. Since the length of the unit in the y direction is the same, the speed difference between the units is not large, and for convenience of calculation, it is only necessary to sum and average the wave speed of each unit to obtain the overall wave speed of the specimen.

The prepared specimens were tested for longitudinal wave speed, as shown in Table 3.4. The average longitudinal wave speed of the original specimen was measured to be 3571 m/s, and the numerical simulation obtained the longitudinal wave speed of the original specimen to be 3577 m/s. In order to correct the reduction coefficient K_0, numerical software was used to simulate the longitudinal sound wave speed of the rock after complete damage under different reduction coefficients. When the reduction coefficient is 0.92, the wave speed obtained by simulation is 1998 m/s, and the longitudinal sound wave propagation speed measured after the specimen is completely damaged under uniaxial loading is 2001 m/s. The two values are closest, and the preset elastic modulus reduction parameter $K_0 = 0.92$ is more reasonable for simulating the reduction of the unit elastic modulus during the damage process of the gypsum specimen, and can reflect the damage change of the gypsum specimen during the loading process.

3) Comparison of strain energy damage model with laboratory results

(1) Comparison of numerical simulation with uniaxial loading test results

The gypsum specimen S-2 was subjected to a uniaxial loading test. The experimental loading rate was 0.5 mm/min. During the experiment, real-time monitoring was carried out using the PAC16 channel acoustic emission device, with the emission threshold set at 43 dB, frequency at 10 kHz–2.1 MHz, and sampling frequency at 1 MHz. To ensure the effectiveness of the experiment, six probes were used for

Table 3.4 Fitting situation of specimen wave speed with different reduction coefficients (unit m/s)

Test type	Test results	Simulation results								
		0.82	0.84	0.86	0.88	0.9	0.92	0.94	0.96	0.98
Original specimen	3571	3577	3577	3577	3577	3577	3577	3577	3577	3577
After damage	2001	1919.2	1933.6	1948.1	1962.6	1977.5	1998	2058	2326	2834

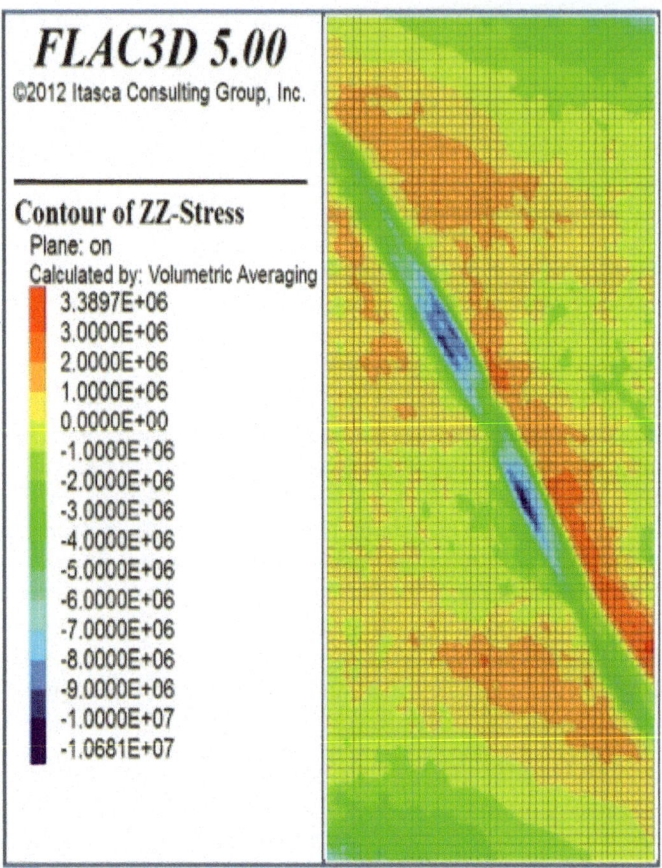

Fig. 3.16 Uniaxial compression SZZ direction force cloud diagram

monitoring, with a preamplification gain of 40 dB, recording the acoustic emission event energy of the specimen loading.

As the stress increases, the strain of the specimen increases, and the local stress distribution cloud diagram of the fracture is shown in Figs. 3.16 and 3.17. The simulated specimen shows a shear failure close to 45° (Fig. 3.18), which is consistent with the experimental results (Fig. 3.19). Comparing the theoretical curve with the uniaxial compression experimental curve of gypsum rock, the theoretical curve can better reflect the stress–strain changes and acoustic emission characteristics of rock fracture. The simulation results are basically consistent with the actual situation of rock fracture, and the simulation calculation method can be applied to rock fracture analysis.

(2) Comparison of numerical simulation with repeated loading test results

3.2 Research on Coal Body Damage Evolution Characteristics

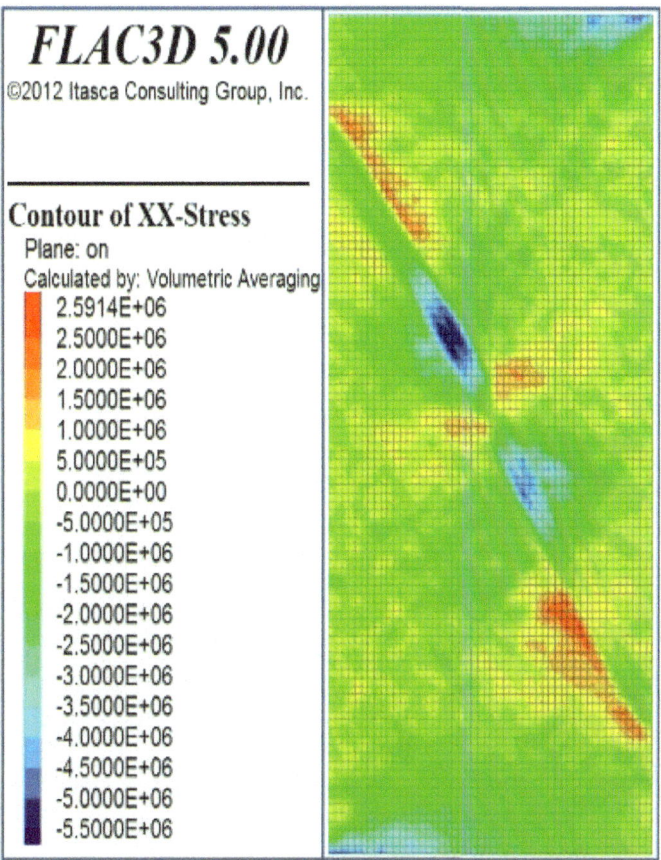

Fig. 3.17 Uniaxial compression SXX direction force cloud diagram

The damage model based on elastic energy accumulates damage strain energy, and has a good simulation effect on the damage analysis of rock specimens under repeated loading conditions. The average uniaxial compressive strength of the specimen $\sigma_t = 35$ MPa. After applying axial loads ($0.3\,\sigma_t, 0.4\,\sigma_t, 0.5\,\sigma_t, 0.6\,\sigma_t$) to specimens S-3, S-4, S-5, and S-6 respectively, they were completely unloaded and then loaded to failure at a speed of 0.5 mm/min.

The parameters of the repeated loading simulation are the same as those of the uniaxial simulation, and the experimental results are compared with the simulation. From Figs. 3.20, 3.21, 3.22, 3.23 and 3.24a, it can be seen that when the specimen is loaded to $0.3\,\sigma_t$ (10.5 MPa) and then unloaded during the simulation experiment, the acoustic emission event energy generated by the first loading is 54,887, which accounts for about 4.8% of the total acoustic emission event energy; when the rock is loaded to $0.4\,\sigma_t$.

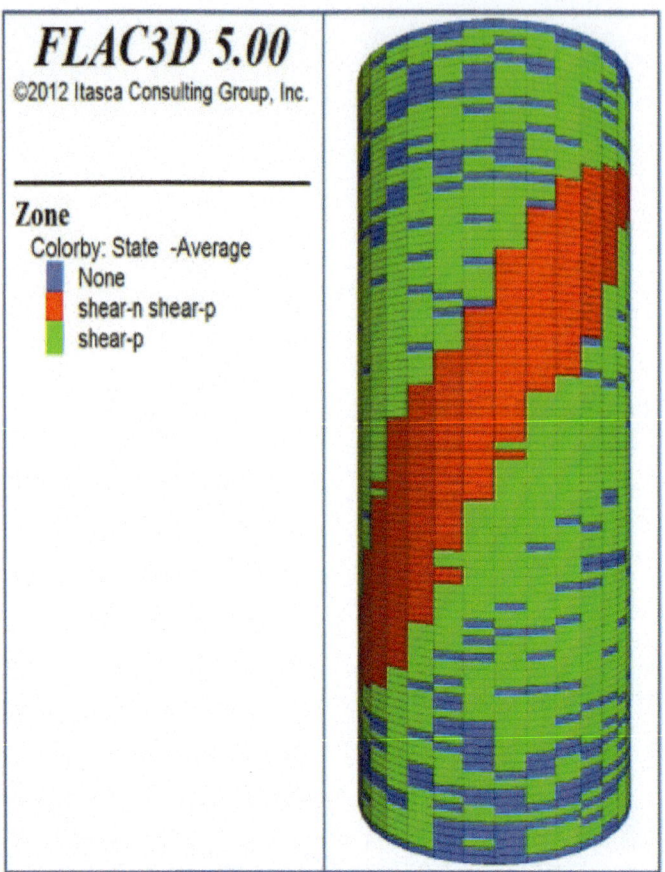

Fig. 3.18 Uniaxial compression simulation result

When the rock is loaded to 0.4 σ_t (14 MPa) and then unloaded, the acoustic emission event energy generated by the initial load is 213,678, accounting for about 18.2% of the total acoustic emission event energy; when the rock is loaded to 0.5 (17.5 MPa) and then unloaded, the acoustic emission event energy generated by the initial load is 391,461, accounting for about 33.4% of the total acoustic emission event energy.

When the rock is loaded to 0.6 σ_t (21 MPa) and then unloaded, the acoustic emission event energy generated by the initial load is 546,150, accounting for about 45.9% of the total acoustic emission event energy.

As can be seen from Figs. 3.20, 3.21, 3.22, 3.23 and 3.24b, when the specimen is loaded to 0.3 σ_t (10.5 MPa) in the laboratory experiment and then unloaded, the acoustic emission event energy generated by the initial load is 49, accounting for about 0.64% of the total acoustic emission event energy; when the rock is loaded to 0.4 σ_t (14 MPa) and then unloaded, the acoustic emission event energy generated by

3.2 Research on Coal Body Damage Evolution Characteristics

Fig. 3.19 Uniaxial compression test result of the specimen

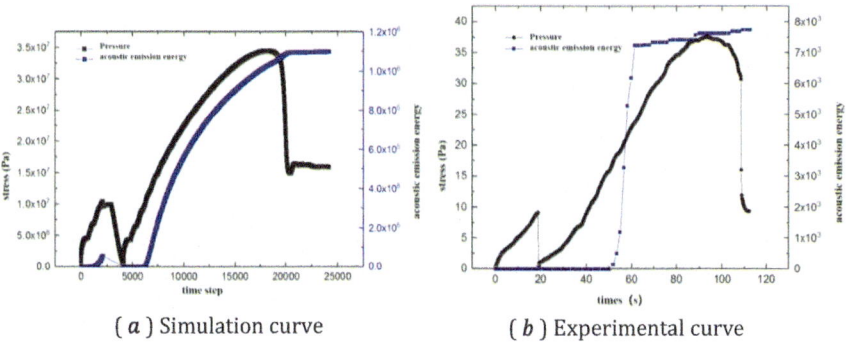

(*a*) Simulation curve (*b*) Experimental curve

Fig. 3.20 Load to 30%σt unload stress-acoustic emission curve

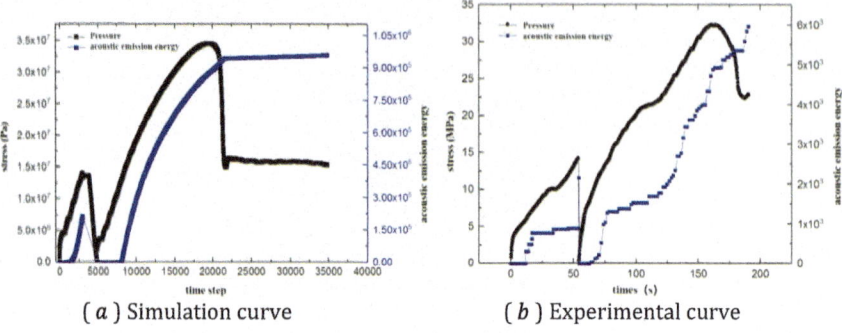

(a) Simulation curve (b) Experimental curve

Fig. 3.21 Load to 40%σt unload stress-acoustic emission curve

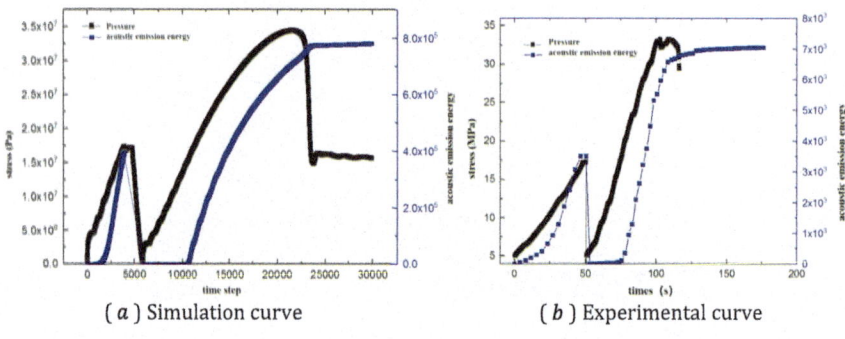

(a) Simulation curve (b) Experimental curve

Fig. 3.22 Load to 50%σt unload stress-acoustic emission curve

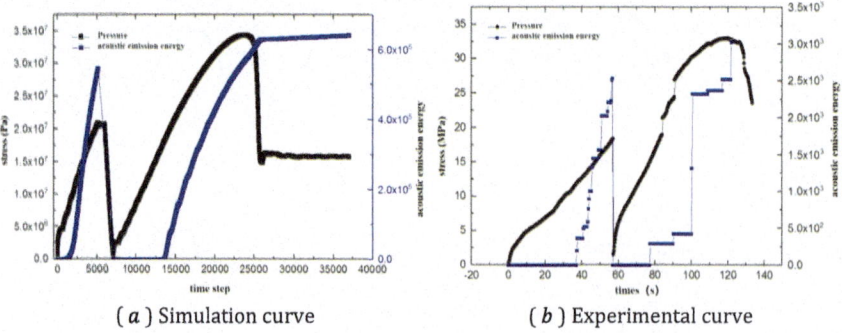

(a) Simulation curve (b) Experimental curve

Fig. 3.23 Load to 60%σt unload stress-acoustic emission curve

3.2 Research on Coal Body Damage Evolution Characteristics

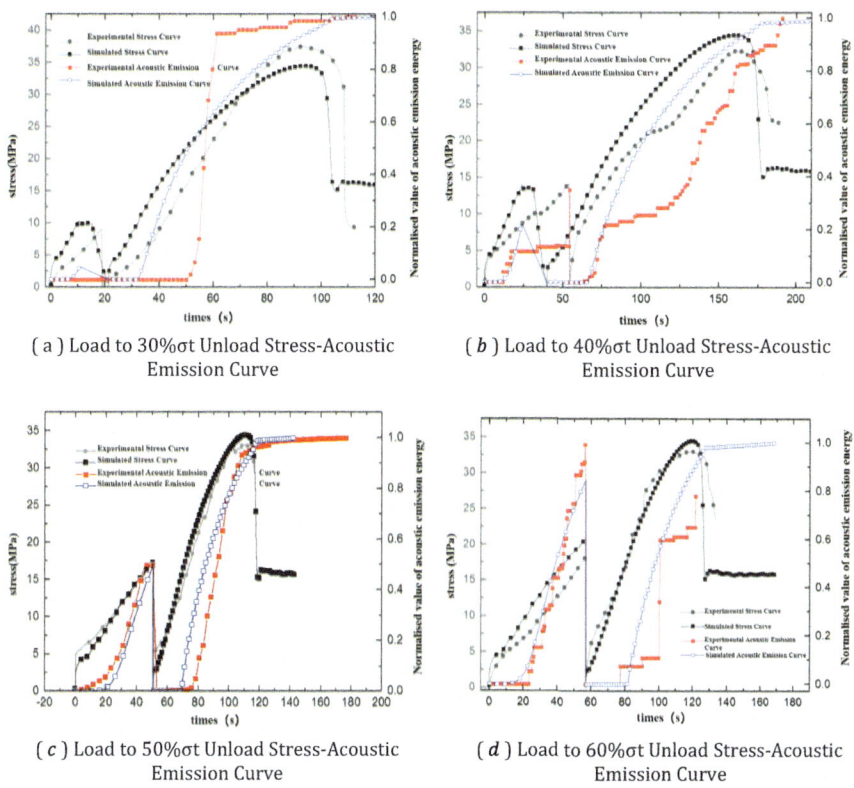

Fig. 3.24 Repeated loading simulation and experimental results

the initial load is 2154, accounting for about 26.52% of the total acoustic emission event energy; when the rock is loaded to 0.5 σ_t (17.5 MPa) and then unloaded, the acoustic emission event energy generated by the initial load accounts for about 33.23% of the total acoustic emission event energy; when the rock is loaded to 0.6 σ_t (21 MPa) and then unloaded, the acoustic emission event energy generated by the initial load is 2537, accounting for about 45.63% of the total acoustic emission event energy.

Both the experimental and simulation results show a clear Kaiser effect of acoustic emission (Fig. 3.24), that is, after reloading, the stress borne by the specimen is less than the maximum stress at unloading, and the specimen will not generate acoustic emission signals. The greater the force of the initial load, the higher the degree of damage to the rock, and the percentage of the acoustic emission energy generated by the initial load in the total acoustic emission energy is correspondingly larger; the percentage of the acoustic emission energy generated in the reloading stage in the total acoustic emission energy will relatively decrease. The proposed strain energy damage constitutive model can reflect the stress and acoustic emission characteristics of gypsum rock under repeated loading and analyze the impact of loading history on rock damage, making this model applicable to the solution of complex mine pillar engineering problems.

3.3 Study on the Spatiotemporal Evolution Law of Mining Stress Field

The mining process is constantly linked to the mining stress, which is the result of the redistribution of the original rock stress caused by mining. The formation and stability of the mining stress field is a development process related to the mining conditions and is directly related to the development and stability of the corresponding rock layer movement. Geostress and mining stress are the fundamental driving forces of coal mining rock layer disasters, and the biggest feature of deep coal mining is that the coal body is in a high original rock stress state before coal resource mining.

In most cases, the coal rock body in the mining field remains stable under the action of three-way stress, but with the mining of the working face, the stress in the vertical direction of the exposed mining space surface quickly drops to atmospheric pressure, and the gas pressure and water pressure in the coal rock body are also the same. The ratio of vertical stress and horizontal stress borne by the coal rock body gradually increases, resulting in stress concentration, forming a large stress gradient in the surrounding rock, making the coal rock body more and more easily destroyed. When the coal rock body is not enough to resist the deformation of the coal body in the stress concentration zone, or when a large amount of energy is released in a short time due to external disturbance stress concentration zone coal body, the working face coal body will become unstable and destroyed, which will then cause surrounding rock movement, rock layer deformation, rock layer structure instability and destruction leading to various disasters. Mining may cause coal seam roof and floor collapse and destruction, support damage, slab roof fall, drumming, and other general mining pressure phenomena, and may also cause impact mining pressure, large area roof pressure, rock burst, mine earthquake, coal and gas outburst, surface subsidence and other large mining dynamic phenomena.

3.3.1 Mining Stress Distribution Characteristics

In front of the working face, when the near coal wall behind the cutting eye changes from a three-way stress state to a two-way (partially one-way) stress state, a stress increase zone is generated within a certain range in front of the coal wall. When the stress exceeds the strength limit of the coal body, the coal body breaks and gradually develops towards the deep, forming the so-called stress reduction zone—stress

3.3 Study on the Spatiotemporal Evolution Law of Mining Stress Field

increase zone—original rock stress zone in front of the working face mining stress impact zone. This mining stress moves forward continuously with the advancement of the working face, and the coal rock body breaking process and breaking range are related to the stress change process and range.

Studies have shown that for different mining depths and coal rock strength conditions, there are three situations of single elasticity, elastic–plastic distribution, and the appearance of internal stress fields in the mining stress distribution on the coal rock body around the mining face.

For a single elastic distribution (as shown in Fig. 3.25a), the stress peak is located at the edge of the coal rock body, and decreases according to the negative exponential curve rule with the increase of the distance from the coal wall. In the entire stress distribution range starting from the coal wall, the coal rock is in an elastic compression state. If the entire process of coal seam destruction is expressed by the full stress–strain curve of coal without impact tendency, then the coal seam in this range is in the elastic deformation stage, that is, the AB stage in Fig. 3.26, and the pressure it bears is proportional to its elastic compression deformation.

Conditions for single elastic distribution: a. The mining depth is shallow, and the initial stress value of the original rock is small; b. The coal rock structure is dense and hard, with high compressive strength; c. The mining (advancement) space is small, and after mining disturbance, the degree of stress concentration is low. At this time,

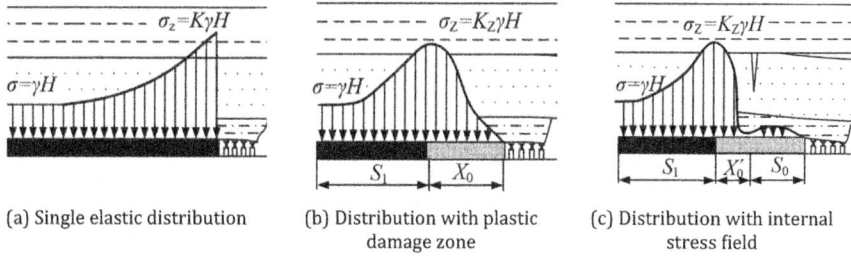

(a) Single elastic distribution

(b) Distribution with plastic damage zone

(c) Distribution with internal stress field

Fig. 3.25 Mining stress distribution characteristics. X_0, X_0' is the plastic zone; S_0 is the internal stress field; S_1 is the elastic zone

Fig. 3.26 Full stress–strain curve of coal without impact tendency

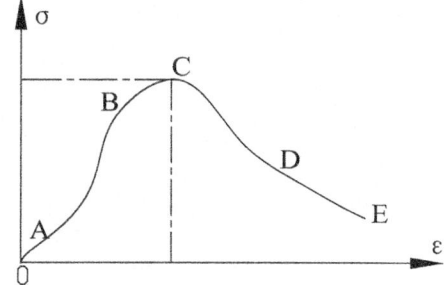

no damage or destruction has occurred at the edge of the coal rock, and the overlying rock layer has not rotated or slipped between layers, and each rock beam can only break at the coal rock wall.

For the distribution with a plastic zone (as shown in Fig. 3.25b), this distribution is composed of a plastic zone X_0 and an elastic zone S_1. The stress peak is at the junction of the elastic zone and the plastic zone, where the elastic zone S_1. The coal rock is in a state of elastic compression deformation, and the pressure values of each part (from the junction of the elastic zone and the plastic zone to the inside of the surrounding rock) are directly proportional to the amount of coal rock deformation. The pressure distribution is a curve with a peak at the junction of elasticity and plasticity and gradually decreases to the original stress value as it moves inward. The interior of the elastic zone is in the original stress state; the coal rock body in the plastic zone is damaged and is in the CDE section of the full stress–strain curve (i.e., the plastic flow stage), and its pressure distribution is a curve that gradually rises from the coal rock wall to the junction of the elastic and plastic zones.

The conditions for the appearance of the plastic zone distribution are: a. The mining depth is deep, and the initial stress value of the original rock is large; b. The coal seam structure is loose and brittle, with low compressive strength; c. The mining (advancing) space is large, and after mining disturbance, the degree of stress concentration is high. Within the range of the plastic zone, the coal rock body has been damaged, its mechanical properties have significantly decreased, and its mechanical bearing capacity is extremely unstable. Therefore, when the overlying rock beam fractures step by step (loses its bearing capacity), the corresponding coal rock area's compression degree gradually increases, the coal rock layer's side is constantly aggravated, and the coal rock mining stress will continuously redistribute, and the peak stress will continuously transfer to the inside of the coal rock body. The main feature of the internal stress field distribution is the fracture of the rock beam deep into the plastic zone. The original intact stress field is clearly divided into two parts along the fracture line of the rock beam, as shown in Fig. 3.25c: one part is the internal stress field determined by the weight of the moving rock beam S_0; the other part is the external stress field related to the total weight of the overlying rock layer, including the newly expanded plastic zone X_0' and the elastic zone S_1 two parts. At this time, the size and influence range of the external stress field pressure are directly related to the mining depth, but the size of the internal stress field pressure only depends on the span and thickness of the moving transfer rock beam, and has no direct connection with the mining depth.

The force conditions of the coal body at different positions within the "internal and external stress field" range are shown in Fig. 3.27. Within the "internal stress field" range (the "S_1" range in Fig. 3.27), the source of coal body damage and destruction comes from the fractured rock beam within the "fracture arch". Because the fractured rock beam fractures and rotates back to the mined-out area, it gives the coal body a dynamic load impact, breaking its equilibrium state and causing coal body damage; within the "external stress field" range (the "$S_2 + S_3$" range in Fig. 3.27), the main

3.3 Study on the Spatiotemporal Evolution Law of Mining Stress Field

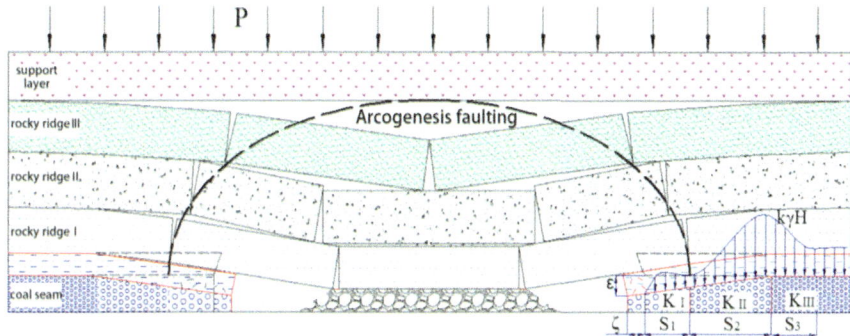

Fig. 3.27 Mechanical model of stable position structure in mining area. ε is the compression amount of the coal wall; ζ is the protrusion amount of the coal wall; $\mathbf{S}_1$ is the range of the "internal stress field"; $\mathbf{S}_2 + \mathbf{S}_3$ is the range of the "external stress field"; $\mathbf{S}_2$ is the range of the pseudo-plastic zone; KI is the porosity of the "internal stress field"; KII is the porosity of the pseudo-plastic zone; KIII is the porosity of the elastic zone

source of coal body force comes from the rock beam action within the "stress arch" range. Due to the continuous fracture and imbalance of the fractured rock beam within the "fracture arch", the distribution of mining stress in the mining area is constantly changing, forming a "stress arch". The force of the rock beam within its range is transmitted layer by layer, acting on the coal body within the range, causing coal body damage. Before the working face is mined, the coal body is in its original equilibrium state. After the mining area advances, under the action of mining stress, the coal body at the coal wall position, i.e., the coal body at the "I" position, is damaged and gradually develops to the "IV" position. At this time, the coal body within the "$S_1 + S_2 + S_3$" range bears the load of the overlying rock layer, and the mining area tends to be stable.

The three types of mining stress mentioned above each have their own conditions of existence. Different coal seams may have different distribution forms under the same mining conditions. Even if the coal seam conditions and mining technical conditions are the same, but the mining depth is different, the different parts of the working face advance, and their distribution composition is often different. Therefore, it is very important to recognize the reasons that affect the existence conditions of various distribution forms for the control of mine pressure, especially for solving the problem of mine pressure control.

The appearance of the internal stress field is based on the existence of the plastic zone. The discriminant formula for the corresponding conditions of the plastic zone in the coal seam is [48]:

$$[H] \geq \frac{\sigma_C}{K\gamma} \qquad (3.33)$$

where, $[H]$ is the critical depth for the formation of the plastic zone under given coal seam conditions; σ_C is the uniaxial compressive strength of the coal seam at a given mining depth. K is the mining stress concentration factor; γ is the rock layer bulk density; H is the mining depth of the coal seam.

From formula 3.33, it can be seen that the larger the mining depth H and the stress concentration factor K, the larger the range of the plastic zone. Given the mining depth and overburden conditions, the mining pressure value of the coal rock mass, including the maximum mining stress concentration factor K and the corresponding mining pressure peak value $K \times \gamma \times H$, also has a certain limit. Therefore, when the mining height is fixed, the maximum range of the plastic zone can also be determined. The higher the strength of the coal rock mass, that is, the higher the uniaxial compressive strength of the coal seam, the smaller the range of the plastic zone under the same mining depth. Under a certain mining depth and given coal seam conditions, the range of the plastic zone is directly proportional to the mining thickness of the coal seam. When mining thick coal seams in layers, the range of the plastic zone depends on the height of the layered mining and the location of the mining.

3.3.2 Study on the Spatiotemporal Characteristics of Mining Stress Evolution

Most existing studies are conducted through numerical simulation or field monitoring, but due to the large number of joints and cracks in the actual coal rock material, its mechanical and acoustic emission characteristics are highly discrete. To overcome this difficulty, the research team independently developed a mining stress test system to study the mechanical characteristics of rock-like materials under different mining stresses, and systematically analyzed the evolution law of mining stress through PFC simulation software.

1. Mining stress evolution test device

(1) Mining stress test system

The research team independently developed a mining stress test system, as shown in Fig. 3.28, and equipped with a 16-channel SH-II acoustic emission device developed by the American Physical Acoustic Corporation (PAC), as shown in Fig. 3.29.

The mining stress tester mainly consists of a vertical loading system, a lateral loading system, a loading main frame, etc., and is a key device for reproducing the mining stress transfer process in the laboratory. The vertical and lateral loading systems of the mining stress tester mainly include loading cylinders, pressure sensors, displacement sensors, return oil valves, overflow valves, safety valves, oil temperature control devices, EDC controllers, etc., which are the main body of the mining stress test system. The side pressure plate is pre-drilled to place the acoustic emission probe. The vertical loading system controls 5 sets of axial loading units, which can independently apply axial stress. Each loading unit is controlled by 2 hydraulic cylinders. During the loading process, the 2 cylinders act at the same time to prevent uneven force during the loading process of a single cylinder, causing the test piece

3.3 Study on the Spatiotemporal Evolution Law of Mining Stress Field

(a) Mining stress tester

(b) Loading unit of mining stress tester

Fig. 3.28 Mining stress tester and its loading unit

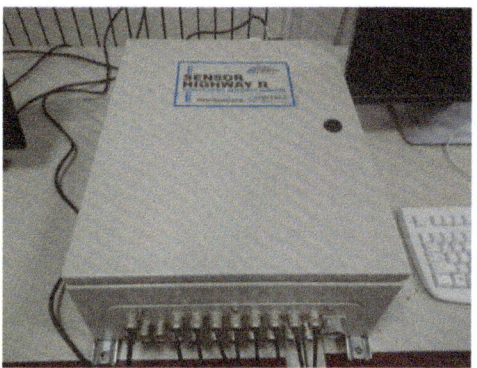
(a) 16-channel SH-II acoustic emission monitoring system

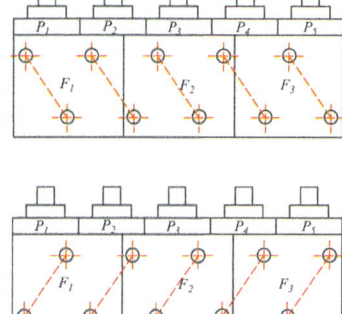
(b) Acoustic emission sensor layout scheme

Fig. 3.29 16-channel SH-II acoustic emission monitoring system and sensor layout scheme

to be sheared and destroyed, and different mining stresses (non-uniform loads) are applied in the vertical direction of the coal rock; the lateral loading system controls 3 sets of horizontal loading units, which are connected by the rear flange method and fixed on the main frame, which can independently apply horizontal stress, and different mining stresses are applied in the horizontal direction of different coal rocks. The main mechanical performance indicators of the vertical and lateral loading system loading units are shown in Table 3.5.

The loading test bench is equipped with removable baffles at the front and back to constrain the displacement of the test piece. The loading test bench can accommodate

Table 3.5 Main mechanical performance indicators of the loading system

Name	Vertical loading unit	Lateral loading unit
Number of loading units	5	1
Size of loading unit/mm	150 mm × 100 mm	167 mm × 500 mm
Load of loading unit/kN	900	500
Stroke of loading cylinder/mm	200	200
Displacement sensor range/mm	300	300
Control accuracy/%	±0.2	±0.5
Load loading rate/kN/s	0.05–100	
Displacement loading rate/mm/min	0.5–100	
Maximum stable time/h	72	

a maximum test piece size of 500 mm (length) × 150 mm (width) × 150 mm (height), and the test piece size can be adjusted as needed.

The mining stress test system uses acoustic emission equipment to monitor the acoustic emission signals generated during the damage and destruction process of coal and rock bodies. This system has the advantages of low threshold, low noise, ultra-fast processing speed, and reliable stability. It can minimize the collection noise, and uses modern digital signal processing technology with 18-bit A/D conversion and enhanced interactive graphical interface to achieve multi-channel high-speed collection, data processing, and real-time analysis of acoustic emission signals during the deformation and destruction process of coal and rock samples.

In order to conveniently monitor the acoustic emission characteristics during the destruction process of coal and rock, a set of 20 mm × 25 mm boreholes are arranged diagonally in each horizontal loading unit to place the acoustic emission probe. The probes in the two horizontal loading units are arranged in an X shape, as shown in Fig. 3.29b. At the same time, small holes are made at the upper part of each borehole to lead out the probe wire. The acoustic emission probe wire can be led out from the small hole, solving the difficulty of monitoring acoustic emission characteristics when applying horizontal load.

(2) Experimental scheme

The mining depth and the advancement speed of the working face are important external factors affecting the stability of the surrounding rock in the mining space. Therefore, based on the mining stress test system, when formulating the test scheme, consider two factors: the initial vertical stress and the stress transfer speed. The stress transfer speed refers to the speed at which the stress acting on the coal and rock body around the mining space is redistributed (transferred) after mining, which is related to the excavation speed. Different initial vertical stresses can quantitatively describe different mining depths. The greater the mining depth, the greater the initial vertical

3.3 Study on the Spatiotemporal Evolution Law of Mining Stress Field

stress; different stress transfer speeds can quantitatively describe different excavation speeds. The faster the mining speed, the faster the stress transfer speed.

The test scheme considers three different initial vertical stresses, with values of 4 MPa, 6 MPa, and 8 MPa, which approximately simulate the mechanical environment of the overlying rock layer with an average density of 2000 kg/m^3, and mining depths of 200 m, 300 m, and 400 m respectively; it considers three different stress transfer speeds, with values of 1 kN/s, 3 kN/s, and 5 kN/s, simulating different mining speeds, and assuming that the stress transfer speed in each area is the same. At the same time, the test scheme considers a confining stress of 4 MPa, simulating the initial horizontal stress of the original rock. The specific experimental scheme is shown in Table 3.6.

Based on the experimental design scheme, the size of the test specimen is 300 mm × 150 mm × 150 mm. In order to quantitatively analyze the mechanical behavior characteristics of the rock-like body under different mining stresses, the rock-like body is now divided into regions. Considering that the experiment considers two major indicators of initial vertical stress and stress transfer speed, it is realized by loading different mining stresses (non-uniform load) through vertical stress, so the rock-like specimen can be divided into regions according to the areas corresponding to the vertical or horizontal loading units of the testing machine. According to the position corresponding to the vertical loading unit, the rock-like specimen is divided into Area A, Area B, and Area C, which correspond to the vertical loading units P1, P2, and P3, respectively. The specimen and its regional division method are shown in Fig. 3.30, and the width-to-height ratio of each region after division is 2:3. At the same time, in order to analyze the mechanical properties of the rock-like under uniaxial compression conditions, three specimens with a size of 100 mm × 100 mm × 200 mm were specially made.

The specific operation steps of the mining stress simulation experiment are as follows:

Step one: Preparation. Place the acoustic emission probe (model R3α) in the probe hole of the horizontal loading unit, so that it corresponds to the positions of Area A, Area B, and Area C. After the arrangement is completed, apply an appropriate lubricant (Vaseline) on the surface of the probe; then run the testing machine, measure the operating noise of the testing machine, and set the acoustic emission monitoring threshold value to 40 db. Place the specimen on the test bench. In order to reduce the friction effect between the loading unit and the end of the specimen during the

Table 3.6 Experimental scheme

Test number	Initial vertical stress/ MPa	Stress transfer speed/ kN/s	Test number	Initial vertical stress/ MPa	Test number	Stress transfer speed/ kN/s	Initial vertical stress/ MPa	Stress transfer speed/ kN/s
1	4	1	4	4	3	7	4	5
2	6	1	5	6	3	8	6	5
3	8	1	6	8	3	9	8	5

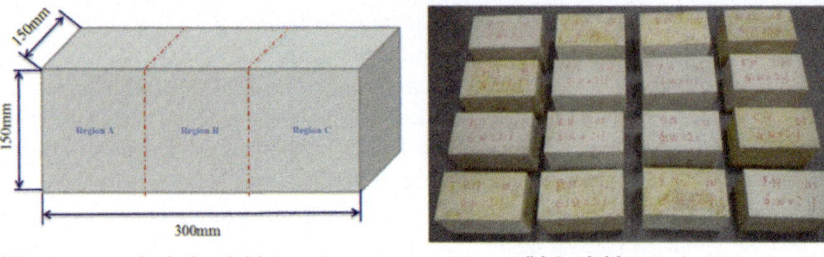

(a) Division method of rock-like specimen area   (b) Rock-like specimen

Fig. 3.30 Division method of rock-like specimen area

test, an appropriate lubricant should be evenly applied to the surface of the specimen before the test;

Step two: Apply pre-stress. Apply initial horizontal stresses F1 and F2 to 4 MPa through displacement control mode (0.5 mm/min), simulating the initial horizontal stress state of engineering coal rock before mining, and then keep it unchanged until the end of the test. Apply initial vertical stresses P1, P2, and P3 to the predetermined values (4, 6, 8 MPa) to Areas A, B, and C of the specimen through synchronous displacement control mode (0.5 mm/min), simulating the initial vertical stress state of engineering coal rock before mining. Area C mainly serves to reduce or eliminate the boundary effect of the test;

Step three: Test process. Conduct a stress transfer test on Area A through force control mode, simulating the stress concentration and unloading failure process near the coal wall after the engineering coal rock is excavated (assumed to be within the range of Area A of the specimen), set the stress transfer speed v (1, 3, 5 kN/s) and stop loading threshold K (0.3, the ratio of the residual strength to the peak strength of the coal rock in each area after the peak), when the loading unit P1 reaches the stop loading threshold, the loading unit P1 becomes the displacement maintenance mode. At this time, the mining stress testing machine will automatically conduct a stress transfer test on Area B, simulating the stress transfer process to the interior (assumed to be within the range of Area B of the specimen) after the engineering coal rock is damaged and a plastic zone is formed (assumed to be within the range of Area A of the specimen), set the stress transfer speed v (1, 3, 5 kN/s) and stop loading threshold K (0.7, because Areas A and C have produced a confining effect on Area B, it is difficult to drop to 30% of the peak, so K = 0.7 is set), stop the test when the loading unit P2 reaches the stop loading threshold.

The above application of pre-stress and test process are set in the stress transfer test module of the mining stress test system, no manual operation is required, and the testing machine automatically completes the stress transfer test. At the same time,

3.3 Study on the Spatiotemporal Evolution Law of Mining Stress Field

the collection of acoustic emission is synchronized with the control of the testing machine, and the monitoring of acoustic emission is real-time and continuous.

(3) Particle flow coal rock compression model

Considering that numerical software can eliminate the weakness of the heterogeneity of real coal rock and has the advantages of high efficiency and speed, the research process also uses the numerical software PFC for simulation and analysis [49–51]. A uniaxial compression model was established using a parallel bonding model to analyze the deformation and failure characteristics of the coal body. Since the simulation experiment using particle flow theory requires setting the microscopic physical and mechanical parameters that characterize the particles and bonding properties, and these parameters cannot be directly obtained from laboratory experiments. Therefore, the "trial and error method" is used to check the parameters required for the model, and the microscopic parameters are adjusted repeatedly until the requirements are met. Through repeated checks and comparisons with the "trial and error method", it is believed that the microscopic physical and mechanical parameters in Table 3.7 are closer to the macroscopic mechanical parameters of the real coal body. After verification, the stress–strain curve (see Fig. 3.31) and the final failure characteristics (see Fig. 3.32) of the particle flow model are in good agreement with the laboratory experiments.

(4) Simulation modeling

The model uses a non-standard size of 90 mm × 40 mm for modeling (see Fig. 3.33). In order to analyze the spatiotemporal rules of coal rock stress evolution during the complete mining dynamics process, three measurement circles with a radius of 15 mm are arranged inside the model (the total monitoring area accounts for 58.9% of the total model area). The measurement centers are A (−30 mm, 0), B (0, 0) and C (30 mm, 0). During the model loading process, the stress (axial) changes in the measurement circle are recorded through the fish language function.

2. Analysis of mining stress evolution characteristics

(1) Rock-like stress–strain characteristics

Table 3.7 Physical and mechanical parameters of coal rock

Parameter	Value	Parameter	Value
Minimum particle size/mm	0.3	Porosity	0.1
Particle size ratio	1.66	Friction coefficient	0.46
Density/(kg/m^3)	1800	Parallel bonding tensile strength/MPa	10
Parallel bonding deformation modulus/ GPa	12	Parallel bonding cohesion/MPa	16

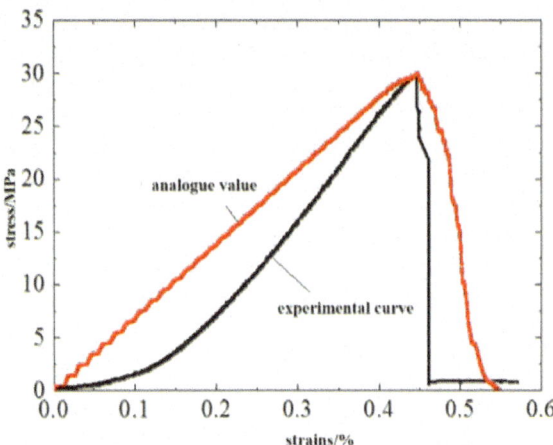

Fig. 3.31 Stress–strain curve of coal rock

Fig. 3.32 Failure mode of coal rock

Before the mining stress test, a uniaxial compression test is first performed on a specimen with a size of 200 mm × 100 mm × 100 mm using the single test mode of the testing machine to measure the uniaxial compressive strength and elastic modulus of the rock-like specimen. The test uses displacement control mode, the

3.3 Study on the Spatiotemporal Evolution Law of Mining Stress Field

Fig. 3.33 Particle flow coal rock compression model

loading speed is set to 0.5 mm/min, and the uniaxial compressive strength of the rock-like specimen is determined to be 15.86 MPa and the elastic modulus is 4.17 GPa through the verification and comparison of three uniaxial compression tests. The stress–strain curve and failure mode are shown in Fig. 3.34.

Figure 3.35 shows the stress–strain curves of different regions of rock-like materials under different mining stress test schemes, and Table 3.8 shows the peak strength and peak strain of regions A and B of rock-like materials under different test schemes. Since the test scheme considers different initial vertical stresses, the rock-like compaction closure stage before the initial vertical stress is ignored, and the data is directly processed from the elastic stage of the rock-like stress–strain curve. In the elastic stage, the stress–strain curves of rock-like regions A and B overlap, which is consistent with the stress evolution characteristics within the coal rock, indicating that the mechanics of the elastic stage of the coal rock or rock-like material will not change significantly due to changes in the stress environment. However, after the stress exceeds the yield limit and enters the plastic stage, the mechanical behavior of each region of the rock-like material shows different evolution characteristics. Region

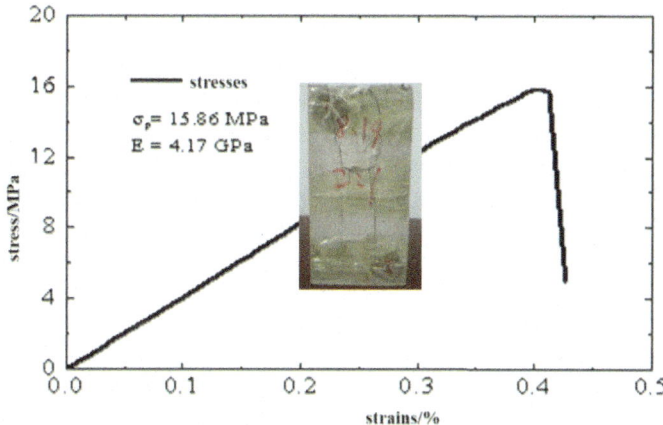

Fig. 3.34 Uniaxial loading stress–strain curve and its failure mode

A will fail and break, while region B can still withstand higher stress. When the initial vertical stress and stress transfer speed are fixed, the peak strength of region B inside the rock-like material is about twice that of region A outside. The main reason is that region A is in a 5-direction stress state when it is damaged and broken, and it is easy to expand and become unstable due to the lack of external constraints. At the same time, region A cannot be completely destroyed when it is unstable, and some or large pieces of rock (rock instability is often caused by shear failure) will remain connected to region B. Therefore, region B is in a 6-direction stress state when it is damaged and broken, and it also includes the constraint of the residual rock blocks of region A, which shows that the constraint of the external residual rock blocks can improve the mechanical bearing capacity of the internal rock. In this experiment, when the volume of the external rock is equal to the volume of the internal rock and the width-to-height ratio is 2:3, the constraint existing after the external rock is broken can increase the strength of the internal rock by approximately 2 times.

In addition, due to the constraint of the external residual rock blocks, the peak elongation of the internal region B is larger, and the deformation amount at the moment of reaching the peak strength is larger, which is several times that of region A. After the peak, the stress–strain change characteristics of region A or region B are different from those under traditional uniaxial compression test conditions. The residual stress after uniaxial compression peak drops rapidly (Fig. 3.34), but the stress drop speed of regions A and B is relatively small due to the effect of non-uniform constraints around, showing different change characteristics under different mining stresses. For region C, because it is always in the initial vertical stress stage, that is, the elastic stage, it will not cause damage, only a small amount of creep deformation.

(2) The impact of initial vertical stress on the mechanical properties of coal rock

Figure 3.36 shows the change characteristics of peak strength in rock-like areas A and B under different initial vertical stress conditions. As can be seen from the figure, when the stress transfer speed is constant, whether it is area A or area B, as the initial vertical stress increases, the peak strength of the rock decreases. According to the generalized Hooke's law, when the initial horizontal stress is constant, the greater the initial vertical stress, the greater the reaction force caused by the initial vertical stress and horizontal stress between regions (region B to region A, region C to region B), which promotes the damage and destruction of the adjacent area, causing the adjacent area to intensify its damage during the gradual compression process, and the compressive strength decreases. In addition, it can be found that under the same stress transfer speed conditions, with the increase of initial vertical stress, the decreasing slope of the peak strength curve of area A and area B is different. When the stress transfer speed is 1 kN/s, the slopes of the peak strength curves of area A and area B are -0.255 and -1.661 respectively; when the stress transfer speed is 3 kN/s, the slopes of the peak strength curves of area A and area B are -0.530 and -0.203 respectively; when the stress transfer speed is 5 kN/s, the slopes of the peak strength curves of area A and area B are -0.504 and -0.491 respectively. From this, it can be seen that the initial vertical stress has different degrees of impact on the strength of area A and area B.

3.3 Study on the Spatiotemporal Evolution Law of Mining Stress Field

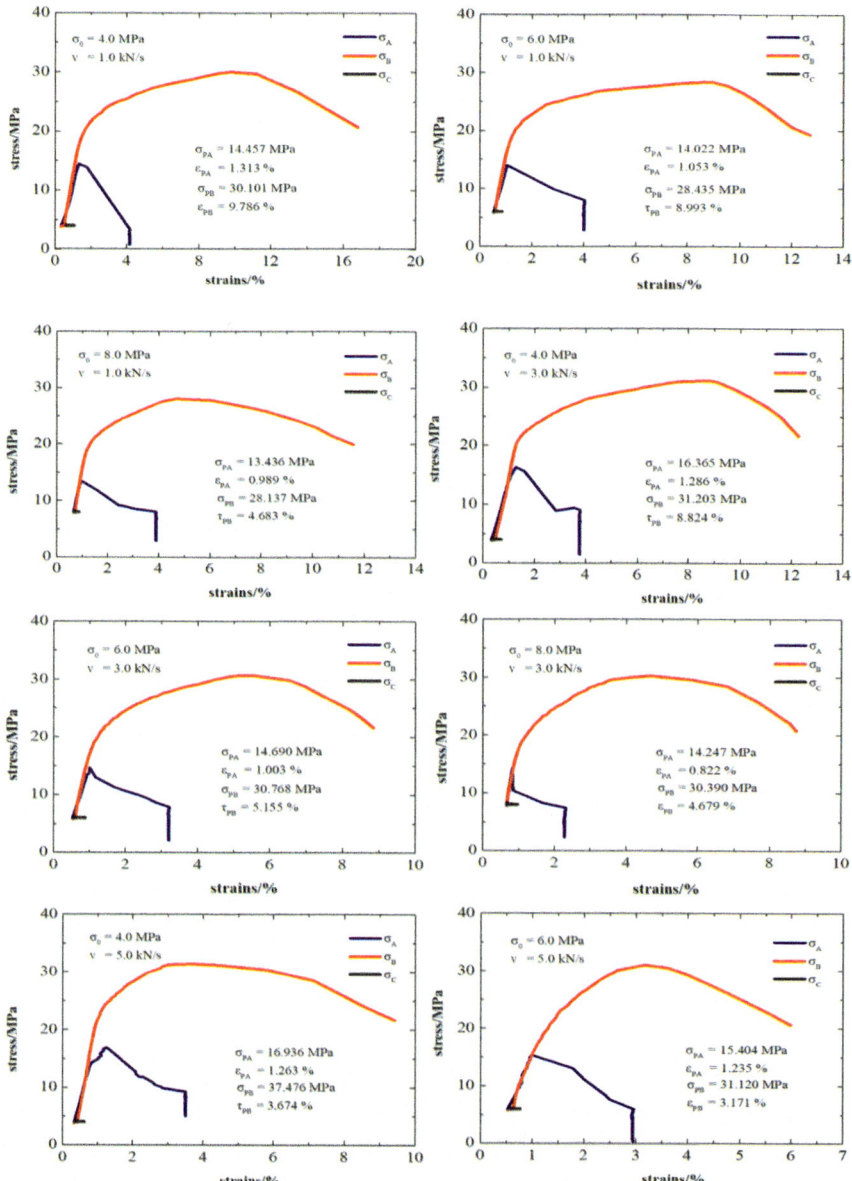

Fig. 3.35 Stress–strain curves of different regions of rock-like materials

Table 3.8 Peak strength and peak strain of regions A and B of rock-like materials

Test number	Peak strength (MPa)		Peak strain (%)	
	Region A	Region B	Region A	Region B
1	14.457	30.101	1.313	9.786
Test number	Peak strength (MPa)		Peak strain (%)	
	Region A	Region B	Region A	Region B
2	14.022	28.435	1.053	8.993
3	13.436	28.137	0.989	4.683
4	16.365	31.203	1.286	8.824
5	14.690	30.708	1.003	5.155
6	14.247	30.390	0.822	4.679
7	16.936	37.476	1.263	3.674
8	15.404	31.120	0.981	3.171
9	14.919	30.834	0.814	2.837

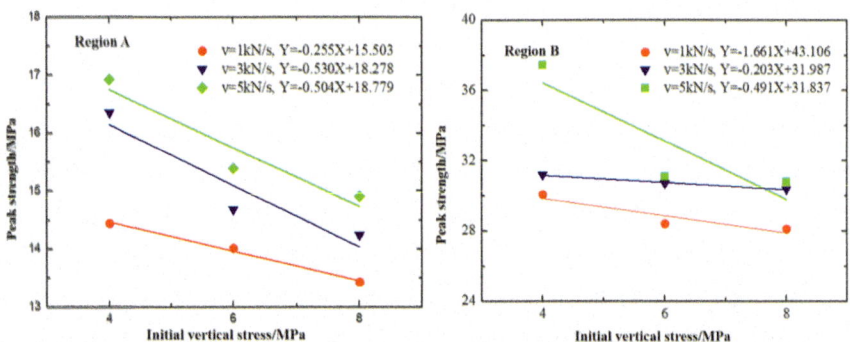

Fig. 3.36 Peak strength of area A and area B under different initial vertical stress conditions

When the stress transfer speed is 1 kN/s, when the initial vertical stress increases from 4 to 6 MPa, the peak strength of area A decreases by 3.0%, and area B decreases by 5.6%. When the initial vertical stress increases from 6 to 8 MPa, the peak strength of area A decreases by 5.9%, and area B decreases by 1.0%. With the increase of initial vertical stress, the decrease of peak stress in area A gradually increases, and the decrease of peak stress in area B gradually decreases, that is, the sensitivity of peak strength in area A increases with the increase of initial vertical stress, while that in area B decreases; When the stress transfer speed is 3 kN/s, when the initial vertical stress increases from 4 to 6 MPa, the peak strength of area A decreases by 10.2%, and area B decreases by 1.6%. When the initial vertical stress increases from 6 to 8 MPa, the peak strength of area A decreases by 3.0%, and area B decreases by 1.0%. With the increase of initial vertical stress, the sensitivity of peak strength in both area A and area B increases, but the increase is different, with area A being

3.3 Study on the Spatiotemporal Evolution Law of Mining Stress Field

greater than area B; When the stress transfer speed is 5 kN/s, when the initial vertical stress increases from 4 to 6 MPa, the peak strength of area A decreases by 9.0%, and area B decreases by 17.0%. When the initial vertical stress increases from 6 to 8 MPa, the peak strength of area A decreases by 4.9%, and area B decreases by 1.1%. With the increase of initial vertical stress, the sensitivity of peak strength in both area A and area B decreases, but the decrease is different, with area B being greater than area A.

Figure 3.37 shows the peak strain characteristics of pseudo-rock regions A and B under different initial vertical stress conditions. As can be seen from the figure, when the stress transfer speed is constant, whether it is region A or region B, as the initial vertical stress increases, the peak deformation of the rock decreases. The main reason for this phenomenon is that the greater the initial vertical stress, the greater the reaction force between regions caused by the initial vertical stress and horizontal stress, the higher the damage and destruction of the adjacent area, the shorter the deformation time of the pseudo-rock from the initial vertical stress stage to the peak strength stage, and the pseudo-rock is damaged and destroyed under smaller deformation. In addition, it can be found that under the same stress transfer speed conditions, with the increase of initial vertical stress, the decreasing slope of the peak deformation curve of region A and region B is also different. When the stress transfer speed is 1 kN/s, the slopes of the peak strain curves of region A and region B are -0.081 and -1.226 respectively; when the stress transfer speed is 3 kN/s, the slopes of the peak strain curves of region A and region B are -0.116 and -0.136 respectively; when the stress transfer speed is 5 kN/s, the slopes of the peak strain curves of region A and region B are -0.112 and -0.209 respectively. From this, it can be seen that the initial vertical stress has different effects on the deformation of region A and region B.

When the stress transfer speed is 1 kN/s, when the initial vertical stress increases from 4 to 6 MPa, the peak strain of region A decreases by 19.8%, and region B decreases by 8.1%. When the initial vertical stress increases from 6 to 8 MPa, the peak strain of region A decreases by 6.0%, and region B decreases by 48.0%. With the increase of initial vertical stress, the decrease of peak strain in region A gradually

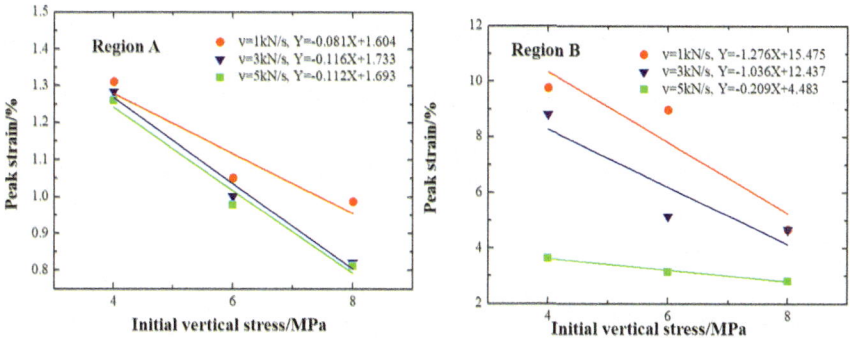

Fig. 3.37 Peak strain of region A and region B under different initial vertical stress conditions

decreases, and the decrease of peak strain in region B gradually increases, that is, the sensitivity of peak strain in region A to the increase of initial vertical stress decreases, while that in region B increases. When the stress transfer speed is 3 kN/s, when the initial vertical stress increases from 4 to 6 MPa, the peak strain of region A decreases by 22.0%, and region B decreases by 42.0%. When the initial vertical stress increases from 6 to 8 MPa, the peak strain of region A decreases by 18.0%, and region B decreases by 9.2%. With the increase of initial vertical stress, the sensitivity of peak strain in both region A and region B decreases, but the decrease in region B is greater than that in region A. When the stress transfer speed is 5 kN/s, when the initial vertical stress increases from 4 to 6 MPa, the peak strain of region A decreases by 22.0%, and region B decreases by 13.7%. When the initial vertical stress increases from 6 to 8 MPa, the peak strain of region A decreases by 17.0%, and region B decreases by 10.5%. With the increase of initial vertical stress, the sensitivity of peak strain in both region A and region B decreases, but the decrease in region A is greater than that in region B.

Overall, the initial vertical stress affects the peak strength and peak deformation of the rock. The greater the initial vertical stress, the lower the peak strength and peak deformation of each region of the pseudo-rock, and the degree of impact on each region is different. This shows that the stress environment has a great impact on the analysis of the mechanical properties of the rock. Although the trend is sometimes close, the quantitative results cannot be equated. The actual engineering problem should use a mechanical environment that is more in line with the actual engineering.

(3) The effect of initial vertical stress on acoustic emission signals

Figure 3.38 shows the acoustic emission impact characteristics of each region of the pseudo-rock under different mining stress conditions. Table 3.9 shows the maximum impact values of region A and region B and the time of the maximum impact. Through analysis, it can be known that because the acoustic emission signal of the initial vertical stress is eliminated, the acoustic emission undergoes four processes during the stress loading and transfer process: the number of impact signals is small, the number of impact signals suddenly increases, then the number of impacts decreases, and then the number of impacts steadily increases. The first two processes correspond to the loading process of region A, and the last two processes correspond to the loading process of region B. The acoustic emission impact characteristics of region A are similar to uniaxial compression, while the acoustic emission impact characteristics of region B are similar to triaxial compression. In addition, it can be seen that the regional characteristics of the maximum impact of the acoustic emission of the pseudo-rock material are also quite obvious. When the initial vertical stress and stress transfer speed are fixed, the maximum impact value of region A is 1.3–2.5 times that of region B. The author believes that this is because region A is brittle failure, and region B is ductile failure.

Figure 3.39 shows the change curve of the maximum impact signal in Area A and Area B under different initial vertical stress conditions. As can be seen from the figure, when the stress transfer speed is constant, whether it is Area A or Area B, as the initial vertical stress increases, the maximum impact signal of the acoustic

3.3 Study on the Spatiotemporal Evolution Law of Mining Stress Field

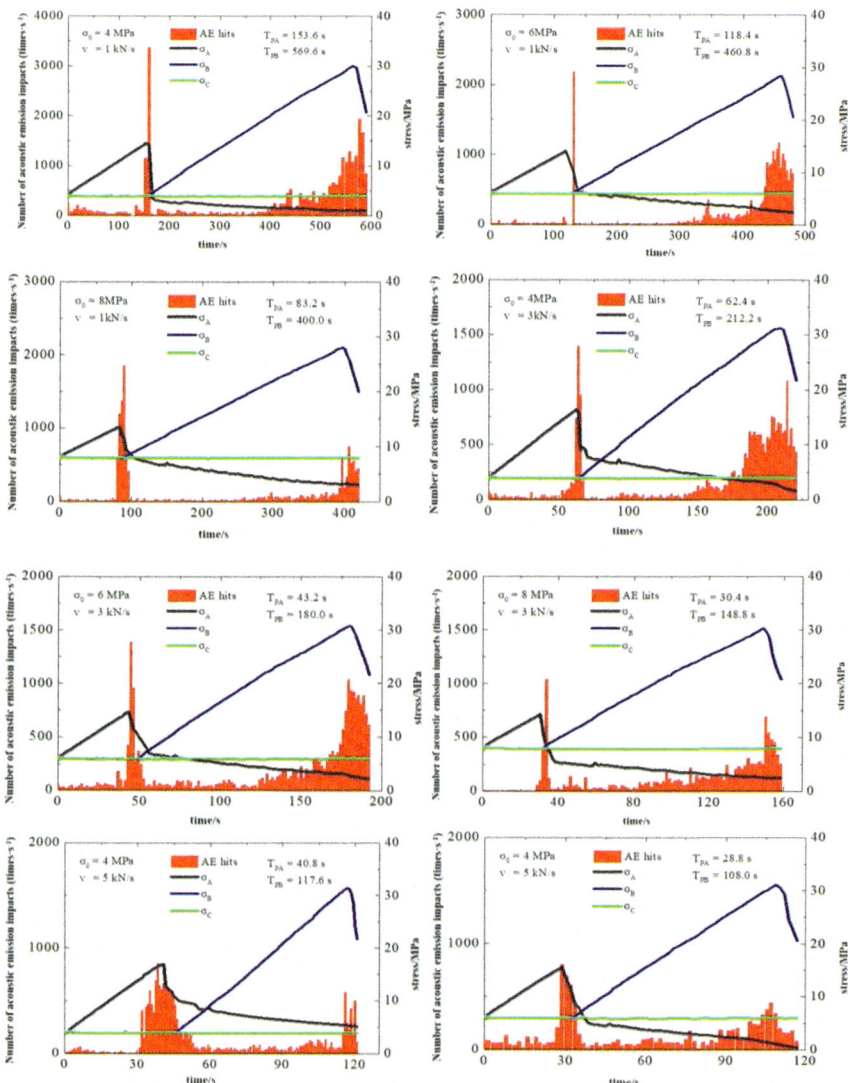

Fig. 3.38 Stress-time-impact curve of each region of pseudo-rock

emission gradually decreases. The generation of the acoustic emission impact signal is due to the squeezing or micro-particle fracture within the pseudo-rock. The greater the initial vertical stress, the better the compaction closure during the initial stress process of the pseudo-rock, and in the later loading process of the pseudo-rock, the smaller the acoustic emission signal generated by particle squeezing. At the same time, the larger initial vertical stress has a stronger induction effect on the damage and destruction of the adjacent area, the destruction in the adjacent area intensifies during

Table 3.9 Maximum acoustic emission impact and occurrence time in pseudo-rock Area A and Area B

Test number	Maximum impact/(times s^{-1})		Time of maximum impact/s	
	Area A	Area B	Area A	Area B
1	3350	1939	153.6	569.6
2	2191	1171	118.4	460.8
3	1858	754	83.2	400.0
4	1411	1080	62.4	212.2
5	1391	1035	43.2	180.0
6	1073	694	30.4	148.8
7	853	579	40.8	117.6
8	801	442	28.8	108.0
9	756	416	24	105.0

the gradual compression process, the destruction process accelerates, the number of destruction faces (blocks) produced in a short time before and after the peak destruction decreases, and the intensity of the acoustic emission signal is smaller. In addition, it can be found that under the same stress transfer speed conditions, with the increase of the initial vertical stress, the slopes of the maximum impact signal curves of Area A and Area B are different. When the stress transfer speed is 1 kN/s, the slopes of the maximum impact signal curves of Area A and Area B are −373.0 and −296.0 respectively; when the stress transfer speed is 3 kN/s, the slopes are −84.5 and −96.5 respectively; when the stress transfer speed is 5 kN/s, the slopes are −24.3 and −40.8 respectively. From this, it can be seen that the initial vertical stress has different effects on the maximum impact signals of Area A and Area B.

When the stress transfer speed is 1 kN/s, the initial vertical stress increases from 4 to 6 MPa, the maximum impact signal in region A decreases by 34.6%, and in region B it decreases by 39.6%. When the initial vertical stress increases from 6 to 8 MPa, the maximum impact signal in region A decreases by 15.2%, and in region B it decreases

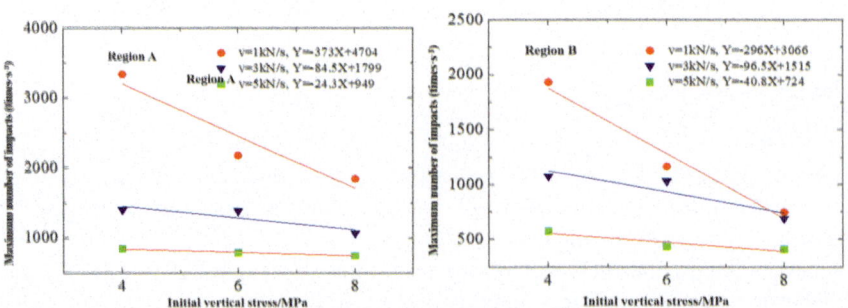

Fig. 3.39 Maximum impact signal of region A and region B under different initial vertical stress conditions

3.3 Study on the Spatiotemporal Evolution Law of Mining Stress Field

by 35.6%. With the increase of the initial vertical stress, the degree of decrease of the maximum impact signal in region A gradually decreases, but the maximum impact signal in region B also gradually decreases. When the stress transfer speed is 3 kN/s, the initial vertical stress increases from 4 to 6 MPa, the maximum impact signal in region A decreases by 1.4%, and in region B it decreases by 4.1%. When the initial vertical stress increases from 6 to 8 MPa, the maximum impact signal in region A decreases by 22.8%, and in region B it decreases by 32.9%. With the increase of the initial vertical stress, the sensitivity of the maximum impact signal in both region A and region B increases, but the increase is different, with region B being greater than region A. When the stress transfer speed is 5 kN/s, the initial vertical stress increases from 4 to 6 MPa, the maximum impact signal in region A decreases by 6.1%, and in region B it decreases by 23.6%. When the initial vertical stress increases from 6 to 8 MPa, the maximum impact signal in region A decreases by 5.6%, and in region B it decreases by 5.8%. With the increase of the initial vertical stress, the sensitivity of the maximum impact signal in both region A and region B decreases, but the decrease is different, with region A only slightly decreasing, and region B decreasing more significantly.

(4) The influence of stress transfer speed on the mechanical properties of coal rock

Figure 3.40 reflects the peak strength change characteristics of region A and region B under different stress transfer speed conditions. As can be seen from the figure, when the initial vertical stress is constant, whether it is region A or region B, the peak strength of the pseudo-rock increases with the increase of stress transfer speed. Different stress transfer speeds have different effects on the degree and form of rock damage. With the increase of stress transfer speed, the stress intensity acting on the rock body at the same time increases, causing the instantaneous damage to the rock body to intensify, and the mode of rock failure changes from shear failure to tensile failure. At the same time, it can be seen that under the same initial vertical stress conditions, with the increase of stress transfer speed, the peak strength of the rock in each region increases at different rates. When the initial vertical stress is 4 MPa, the slope of the peak strain curve of region A and region B is 1.240 and 3.688 respectively; when the initial vertical stress is 6 MPa, the slope of the peak strain curve of region A and region B is 0.691 and 1.343 respectively; when the initial vertical stress is 8 MPa, the slope of the peak strain curve of region A and region B is 0.742 and 1.349 respectively. The effect of stress transfer speed on the peak strength of each region of the pseudo-rock is different.

When the initial vertical stress is 4 MPa, the stress transfer speed increases from 1 to 3 kN/s, the peak strength of region A increases by 11.7%, and region B increases by 3.5%. When the stress transfer speed increases from 3 to 5 kN/s, the peak strength of region A increases by 3.4%, and region B increases by 16.7%. With the increase of stress transfer speed, the increment of peak stress in region A gradually decreases, and the increment of peak stress in region B gradually increases. When the initial vertical stress is 6 MPa, the stress transfer speed increases from 1 to 3 kN/s, the peak strength of region A increases by 4.5%, and region B increases by 7.4%. When the stress transfer speed increases from 3 to 5 kN/s, the peak strength of region

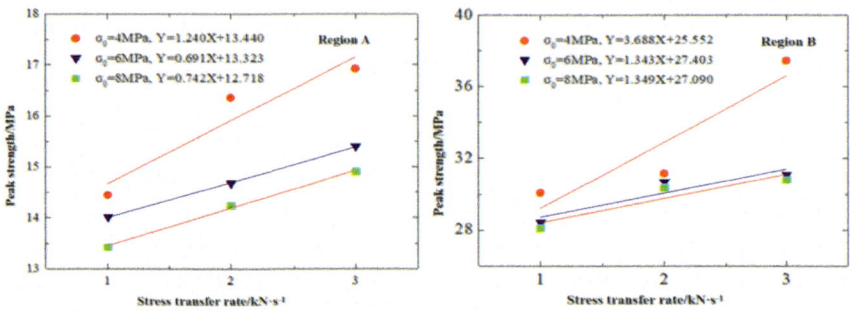

Fig. 3.40 Peak strength of region A and region B under different stress transfer speed conditions

A increases by 4.8%, and region B increases by 1.3%. With the increase of stress transfer speed, the sensitivity of the peak strength of region A slightly increases, while that of region B decreases. When the initial vertical stress is 8 MPa, the stress transfer speed increases from 1 to 3 kN/s, the peak strength of region A increases by 5.7%, and region B increases by 7.4%. When the stress transfer speed increases from 3 to 5 kN/s, the peak strength of region A increases by 4.5%, and region B increases by 1.4%. With the increase of stress transfer speed, the sensitivity of the peak strength of both region A and region B decreases, but the decrease is different, with region B being greater than region A.

As shown in Fig. 3.41, when the initial vertical stress is constant, whether it is region A or region B, the peak strain decreases with the increase of stress transfer speed. This is because the greater the stress transfer speed, the greater the stress acting on the pseudo-rock at the same moment, the greater the instantaneous damage to the pseudo-rock, causing the pseudo-rock to damage and break without significant deformation. At the same time, it can be seen that under the same initial vertical stress conditions, with the increase of stress transfer speed, the peak strain reduction slopes of different regions of the rock are different. When the initial vertical stress is 4 MPa, the peak strain curve slopes of region A and region B are −0.025 and −3.056 respectively; when the initial vertical stress is 6 MPa, the peak strain curve slopes of region A and region B are −0.036 and −2.911 respectively; when the initial vertical stress is 8 MPa, the peak strain curve slopes of region A and region B are −0.088 and −0.923 respectively. The stress transfer speed has different effects on the peak strength of different regions of the pseudo-rock, and the stress transfer speed has different effects on the peak deformation of different regions of the rock.

When the initial vertical stress is 4 MPa, when the stress transfer speed increases from 1 to 3 kN/s, the peak strain of region A decreases by 2.1%, and region B decreases by 9.8%. When the stress transfer speed increases from 3 to 5 kN/s, the peak strength of region A decreases by 1.8%, and region B decreases by 58.4%. With the increase of stress transfer speed, the decrease of peak strain in region A gradually decreases, and the decrease of peak strain in region B increases. When the initial vertical stress is 6 MPa, when the stress transfer speed increases from 1 to 3 kN/s, the peak strain of region A decreases by 4.7%, and region B decreases by

3.3 Study on the Spatiotemporal Evolution Law of Mining Stress Field

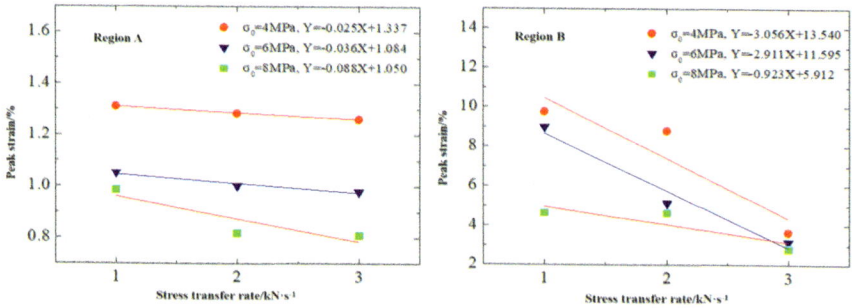

Fig. 3.41 Peak strain of region A and region B under different stress transfer speed conditions

42.7%. When the stress transfer speed increases from 3 to 5 kN/s, the peak strength of region A decreases by 2.2%, and region B decreases by 38.5%. With the increase of stress transfer speed, the sensitivity of peak strain in both region A and region B decreases, but the decrease is different, with region B greater than region A. When the initial vertical stress is 8 MPa, when the stress transfer speed increases from 1 to 3 kN/s, the peak strain of region A decreases by 16.8%, and region B decreases by 0.8%. When the stress transfer speed increases from 3 to 5 kN/s, the peak strength of region A decreases by 0.9%, and region B decreases by 39.4%. With the increase of stress transfer speed, the sensitivity of peak deformation in region A drops sharply, while the sensitivity of peak strain in region B increases sharply.

In general, the stress transfer speed affects the peak strength and peak deformation of the rock. The greater the stress transfer speed, the higher the peak strength of each region of the coal rock, but the lower the peak strain. The impact of different stress transfer speeds on different regions of the pseudo-rock is different and closely related to the initial vertical stress.

(5) The effect of stress transfer speed on acoustic emission signals

Figure 3.42 is the change curve of the maximum impact signal in region A and region B under different stress transfer speed conditions. As can be seen from the figure, when the initial vertical stress is constant, whether it is region A or region B, with the increase of stress transfer speed, the maximum impact signal of acoustic emission gradually decreases. The greater the stress transfer speed, the greater the instantaneous damage to the pseudo-rock, the acceleration of the damage and destruction of the pseudo-rock, and the larger the degree of destruction during the peak destruction process. At the same time, by analyzing the maximum impact signals of different regions of the pseudo-rock under the same initial vertical stress conditions, it can be known that under the same vertical stress conditions, with the increase of stress transfer speed, the slopes of the maximum impact signal curves of region A and region B are different. When the initial vertical stress is 4 MPa, the slopes of the maximum

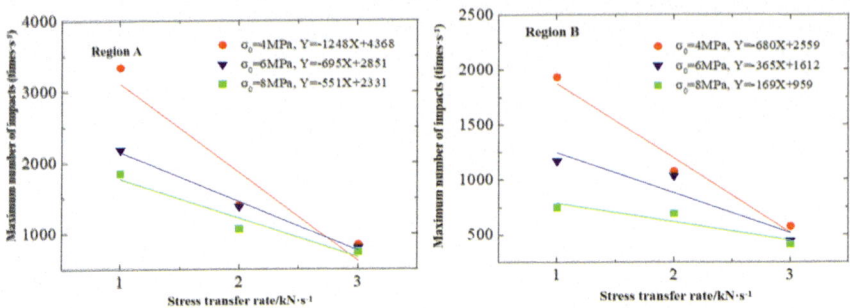

Fig. 3.42 Maximum impact signal of region A and region B under different stress transfer speed conditions

impact signal curves of region A and region B are −1284 and −680 respectively; when the initial vertical stress is 6 MPa, the slopes of the maximum impact signal curves of region A and region B are −695 and −365 respectively; when the initial vertical stress is 8 MPa, the slopes of the maximum impact signal curves of region A and region B are −551 and −169 respectively. The stress transfer speed has different effects on the maximum impact signals of different regions of the rock.

When the initial vertical stress is 4 MPa, the stress transfer speed increases from 1 to 3 kN/s, the maximum impact signal in region A decreases by 57.9%, and region B decreases by 44.3%. When the stress transfer speed increases from 3 kN/s to 5 kN/s, the maximum impact signals in region A and region B decrease by 39.5% and 46.4% respectively. As the stress transfer speed increases, the degree of decrease in the maximum impact signal in region A gradually decreases, but the decrease in region B slightly increases. When the initial vertical stress is 6 MPa, the stress transfer speed increases from 1 kN/s to 3 kN/s, the maximum impact signals in region A and region B decrease by 36.5% and 11.6% respectively. When the stress transfer speed increases from 3 kN/s to 5 kN/s, the maximum impact signals in region A and region B decrease by 42.4% and 57.3% respectively. As the stress transfer speed increases, the sensitivity of the maximum impact signals in both region A and region B increases, with the increase in region B being greater than that in region A. When the initial vertical stress is 8 MPa, the stress transfer speed increases from 1 to 3 kN/s, the maximum impact signal in region A decreases by 42.2%, and region B decreases by 8.0%. When the stress transfer speed increases from 3 to 5 kN/s, the maximum impact signal in region A decreases by 29.5%, and region B decreases by 40.1%. As the stress transfer speed increases, the sensitivity of the maximum impact signal in region A gradually decreases, while the sensitivity in region B sharply increases.

3. Evolutionary characteristics of mining stress

Figure 3.43 is the coal rock compression stress-time step curve. The green curve in the figure represents the change characteristics of the overall average stress of the coal rock (wall loading stress), and the red, blue, and black curves represent the stress change characteristics of the measured areas A, B, and C respectively. Figure 3.44

3.3 Study on the Spatiotemporal Evolution Law of Mining Stress Field

is the coal rock compression stress concentration coefficient-time step curve, where the defined stress concentration coefficient

$$K = \frac{\sigma_i}{\sigma} \tag{3.34}$$

where, σ_i is the stress at a certain moment in the i measured area; σ is the average stress of the entire model at a certain moment.

As can be seen from Fig. 3.43, with the continuous change of coal rock compression stress, the overall and local measurement areas A, B, and C of the coal rock all have the same stress evolution characteristics, all experiencing the initial compression compaction stage, elastic compression stage, plastic deformation stage, and damage residual stage. Different from the traditional coal rock compression test, the concave characteristic of the initial compaction stage of the numerical model is not obvious, which is because the particle balls simulating coal rock in the particle flow program are rigid balls and do not deform. In the coal rock compression compaction stage and elastic deformation stage, the overall and local measurement areas A, B, and C

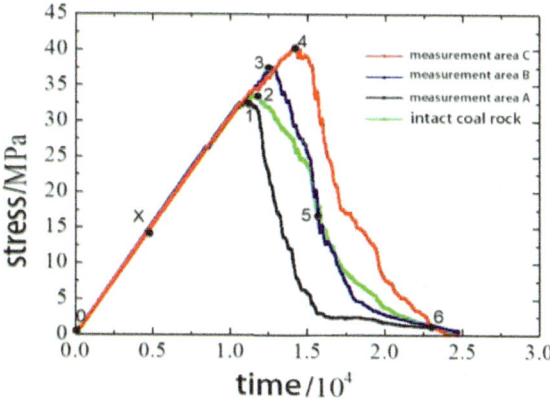

Fig. 3.43 Coal rock compression stress-time step curve

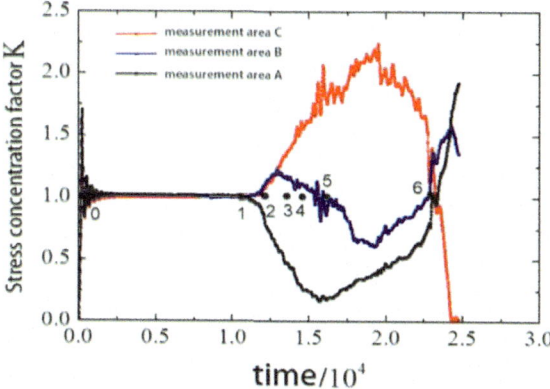

Fig. 3.44 Coal rock compression stress concentration coefficient-time step curve

have the same change characteristics, but they show different mechanical properties in the plastic deformation stage and damage residual stage, which is the key stage of mining stress spatiotemporal movement.

As can be seen from Fig. 3.44, during the compression process of coal rock, the degree of local stress concentration in the coal body has a significant spatiotemporal effect. As the overall force of the coal rock gradually increases (the green curve 0-X-2 point in Fig. 3.43, point X ~ point 2 can reflect the actual engineering coal rock body from the original stress (point X) to the sudden unloading stress concentration process), the stress in the three measurement areas also gradually increases, and the stress concentration coefficient of each monitoring area is approximately 1. However, at a certain t1 time domain before the overall coal sample is subjected to the peak stress (shown as points 1–2 in the figure), the stress concentration coefficient of the external coal rock (measurement area A) begins to decrease, indicating that the coal rock in measurement area A has damage and destruction, leading to a decrease in its own mechanical bearing capacity, and the stress begins to transfer to the interior.

As can be seen from Table 3.10, at key point 1, that is, at the stress peak of region A, the stress values of coal rock regions A, B, C and the intact coal rock are all approximately 32.8 MPa, at time step 11,005; at key point 2, that is, when the intact rock mass reaches the stress peak, the stress values of coal rock regions A, B, C and the intact coal rock are 31.4 MPa, 35.1 MPa, 34.6 MPa and 33.7 MPa respectively, at time step 11,778. The stress of the intact coal rock increased by 0.9 MPa compared to point 1, the stress in measurement area A decreased by 1.4 MPa, the stress in measurement area B increased by 2.3 MPa, and the stress in measurement area C increased by 1.8 MPa. Measurement areas B and C not only have to bear their own increased stress, but also the stress transferred from area A. Subsequently, in the t2 time domain (points 2–3 in the figure), the central coal rock (monitoring area B) reaches the stress peak, but the stress concentration factor is still greater than 1, indicating that the central coal rock can continue to bear part of the overall stress of the coal rock in excess, and it bears more additional stress than measurement area C. Then, the stress continues to transfer to the inside of the coal body, in the t3 time domain (points 3–4 in the figure), the internal coal rock (measurement area C) reaches the stress peak, the stress in measurement areas A and B continues to decrease, but the stress concentration factor in measurement area B is still greater than 1, indicating that it can still bear some of the stress transferred from measurement area A, and the force transferred from area A to the inside is mainly borne by area C, because the stress concentration factor in area C continues to rise. Subsequently, the bearing capacity of each area of the coal rock begins to decrease, in the t4 time domain (points 4–5 in the figure), the stress concentration factor of the central coal rock damage decreases, and the overall damage stress of the coal rock is mainly borne by the internal coal rock. Later (at this time, the nominal stress of the overall coal rock is 46.8% of the peak stress), until the t5 time domain (points 5–6 in the figure) when the intact coal rock completely loses its bearing capacity, the stress concentration factor of the central coal rock damage is less than 1, and the residual damage stress is completely concentrated in the interior of the coal rock.

3.3 Study on the Spatiotemporal Evolution Law of Mining Stress Field

Table 3.10 Stress evolution characteristics of key points 1–5

Time/Step	Region A stress/MPa	Region B stress/MPa	Region C stress/MPa	Intact coal rock stress/MPa
Point 1 at 11,005	32.8	32.8	32.8	32.8
Point 2 at 11,778	31.4	35.1	34.6	33.4
Point 3 at 12,513	23.7	37.6	36.1	32.2
Point 4 at 14,233	10.6	29.7	40.2	26.8
Point 5 at 15,670	3.2	18.3	33.3	18.2

In summary, the damage evolution process of coal rock has obvious spatiotemporal characteristics, and the stress at each spatial position has the following relationship with time:

$$\left(\begin{array}{c} k_{A0} = 1..k_{A4} \leq k_{A3} \leq k_{A2} \leq k_{A1} \leq 1..k_{A5} \leq 1 \\ k_{B0} = 1..k_{B2} \geq k_{B3} \geq k_{B4} \geq k_{B1} \geq 1..k_{A5} \leq 1 \\ k_{C0} = 1..k_{C4} \geq k_{C3} \geq k_{C2} \geq k_{C1} \geq 1..k_{A5} \geq 1 \\ t = 0 \sim t_1 \ldots\ldots\ldots\ldots..t = t_1 \sim t_4 \ldots\ldots\ldots\ldots.t = t_5 \end{array} \right) \quad (3.35)$$

where, k_{ni} is the stress concentration factor of each spatial monitoring area within the i time range, $n = A, B, C$.

Figure 3.45 is a cloud diagram of mining stress evolution. As can be seen from the cloud diagram, within the coal rock measurement area, with the passage of time, the stress cloud diagram experiences a distribution feature from small to large and then to small, and the stress peak areas on the left and right are not symmetrical. The left side shows a uniform increase, while the right side shows a fluctuating evolution. This corresponds to the stress-time curve.

By comparing the cloud diagrams of Area A, Area B, and Area C, it can be seen that within a certain range of coal body, the stress evolution time in the external area of the coal rock (Area A) is the shortest, and the further into the coal rock (Area B or Area C), the longer the stress evolution cycle and the higher the stress concentration. At the same time, it can be seen that whether it is the internal or external area of the coal rock, the characteristics of coal rock mining stress evolution are almost the same. When they exceed their respective limit strengths, the residual damage stress in the internal area is greater than that in the external area.

Fig. 3.45 Mining stress evolution cloud diagram

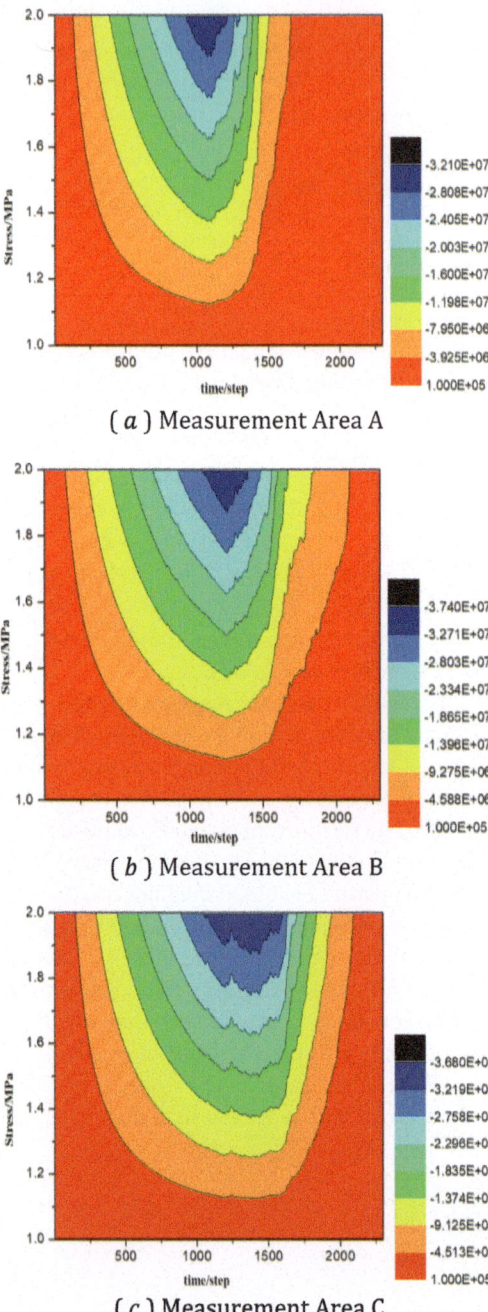

(*a*) Measurement Area A

(*b*) Measurement Area B

(*c*) Measurement Area C

References

1. Zhou H, Xie H, Zuo J (2005) Progress in research on rock mechanical behavior under deep high ground stress. Adv Mech 35(1):91–99
2. Xie H, Zhou H, Xue D et al (2012) Research and thinking on deep coal mining and limit mining depth. J Coal 37(04):535–542
3. He M (2005) Conceptual system of deep and engineering evaluation indicators. Chin J Rock Mech Eng (16):2854–2858
4. Zhang JM, Li QS, Zhang Y et al (2019) Definition and mining response analysis of deep coal mining. J Coal 44(5):1314–1325
5. Xie H, Gao F, Ju Y et al (2015) Quantitative definition and analysis of deep mining. J Coal 40(01):1–10
6. Wang L (1994) Application of ground stress measurement in mining engineering. Seismological Press
7. Brown ET, Hoek E (1978) Trends in relationships between measured in-situ stresses and depth. Int J Rock Mech Min Sci Geomech Abstracts 15(4):211–215
8. Zhang J, Qiu W (2011) Analysis of the influence of geological structure and terrain on N. Barton's rock quality designation Q value. Mod Tunnel Technol 48(06):38–42
9. Shen H, Cheng Y, Wang J et al (2007) Finite element study on the influence of faults on the stress field. Daqing Petrol Geol Dev (02):34–37
10. Wang Y, Cui X, Hu X et al (2012) Study on the stress state of the upper crust of the Chinese mainland based on in-situ stress measurement data. Chin J Geophys 55(09):3016–3027
11. Kang H (2013) Stress distribution characteristics and roadway surrounding rock control technology in deep coal mines. Coal Sci Technol 41(9):12–17
12. Hoek E, Brown ET (1982) Closure of "empirical strength criterion for rock masses." J Geotech Eng Div 108:672–673
13. Cleary M (1989) Effects of depth on rock fracture. In: Manry V, Fourmaintrax D (eds) Rock at great depth. A. A. Balkema, Rotterdam, pp 1153–1163
14. Li S (1996) Concise course on rock mechanics. Coal Industry Publishing House
15. Xie H (2017) Research ideas and expected results of "deep rock mechanics and mining theory". Eng Sci Technol 49(02):1–16
16. Zhao Z, Tian Q, Li L (2005) Study on the fracture mechanism of coal rock mass. J Taiyuan Univ Technol 36(3):260–263
17. Chen Z, Tang C, Fu Y (1998) Numerical simulation of rock micro-fracture damage evolution induced by mutation. Chin J Geotech Eng 20(6):9–15
18. Pan J (2006) The relationship between the deformation and failure mechanism of coal rock under uniaxial compression and its impact tendency. Coal Mine Saf 37(8):1–4
19. Xia M, Han W, Ke F (1995) Statistical mesoscopic damage mechanics and damage evolution induced mutation (I). Adv Mech 25(01):1–14
20. Xia M (1995) Statistical mesoscopic damage mechanics and damage evolution induced mutation (II). Adv Mech 25(02):145–159
21. Liu B, Huang J, Wang Z et al (2009) Study on damage evolution and acoustic emission characteristics of coal rock under uniaxial compression. Chin J Rock Mech Eng 28(1):3234–3238
22. Zou Y (2004) Preliminary exploration of coal rock acoustic emission mechanism. Min Saf Environ Protect 31(1):31–33
23. Yang Y, Wang D, Guo M et al (2014) Study on rock damage characteristics based on triaxial compression acoustic emission test. Chin J Rock Mech Eng 33(1):98–104
24. Xu J, Li S, Tang X et al (2008) Analysis of influencing factors on rock acoustic emission localization test under uniaxial compression. Chin J Rock Mech Eng 27(4):765–772
25. Kong BH, Li Z, Wang EY (2018) Fine characterization of rock thermal damage by acoustic emission technique. J Geophys Eng 15(1):1–10
26. Liu X, Lin J, Yuan Z (1997) Research on the evaluation of material fatigue damage using acoustic emission technology. China Railw Sci 18(4):74–81

27. Song X, Li Z (2011) Study on the acoustic emission characteristics of coal samples under uniaxial compression. Coal Eng 1(11):110–113
28. Ning C, Yu F, Jing L (2011) Experimental study on the acoustic emission characteristics of coal rock under uniaxial compression. Coal Min 16(1):97–100
29. Li H, Gao B, Li H (2015) Study on the macro fracture structure and acoustic emission characteristics of coal rock under uniaxial compression. J Undergr Space Eng 11(3):612–618
30. Gao B, Li H, Liu Y et al (2013) Study on the acoustic emission and fractal characteristics of coal rock under uniaxial compression. J Undergr Space Eng 10(5):19–25
31. Lin QB, Cao P, Li KH et al (2018) Experimental study on acoustic emission characteristics of jointed rock mass by double disc cutter. J Central South Univ 25(2):357–367
32. Zuo J, Pei J, Liu J et al (2011) Acoustic emission behavior and spatio-temporal evolution mechanism during the fracture process of coal rock mass. Chin J Rock Mech Eng 30(8):1564–1570
33. Xie H, Peng R, Ju Y (2004) Analysis of energy dissipation during rock failure process. Chin J Rock Mech Eng 23(21):3565–3570
34. Kachanov LM (1976) Separation failure of composite materials. Polym Mech 12(5):812–815
35. Wen Z, Tian L, Jiang Y et al (2019) Research on the damage constitutive model of heterogeneous rock based on strain energy density. Chin J Rock Mech Eng 38(07):1332–1343
36. Xie H, Ju Y, Li L et al (2008) Energy mechanism of rock deformation and failure process. Chin J Rock Mech Eng (09):1729–1740
37. Gao F, Zhang Z, Liu X (2012) Research on rock burst proneness index based on energy evolution in rock. Disaster Adv 5(4):1367–1371
38. Wen Z, Wang X, Chen L et al (2017) Size effect on acoustic emission characteristics of coal-rock damage evolution. Adv Mater Sci Eng
39. Sun Q, Li S, Feng X et al (2011) Research on numerical simulation method of rock fracture based on strain energy density theory. Rock Soil Mech 32(05):1575–1582
40. Li X, Cao WG, Su YH (2012) A statistical damage constitutive model for softening behavior of rocks. Eng Geol 143:1–17
41. Yang S, Xu W, Wei L et al (2004) Study on statistical constitutive model and experiment of rock damage under uniaxial compression. J Hohai Univ (Nat Sci) (02):200–203
42. Zeng S, Yang S, Zhang X et al (2005) Study on statistical constitutive model and experiment of limestone damage under uniaxial compression. J Univ South China (Sci Technol) (01):69–72+95
43. Yang M, Zhao M, Cao W (2005) Method for determining parameters of rock damage softening statistical constitutive model. J Hydraul Eng (03):345–349
44. Xu T, Tang C, Zhang Z et al (2003) Theory, experiment and numerical simulation of brittle rock deformation and failure under uniaxial compression. J Northeast Univ (01):87–90
45. Zhang X (2010) Analysis of rock damage statistical constitutive model parameters and their critical sensitivity. J Min Saf Eng 27(01):45–50
46. Zhang F (1990) Comparative analysis of field dynamic static method elastic parameter measurement results. Water Resour Hydropower Technol 11:34–39
47. Zhang P, Zhang X, Wang T (2001) Relationship between rock elastic modulus and elastic wave velocity. Chin J Rock Mech Eng 20(6):785–785
48. Yin Z (2007) Research on the characteristics and application of mining overburden damage. Shandong University of Science and Technology, Qingdao
49. Itasca Consulting Group (2008) PFC2D (particle flow code in 2 dimensions) fish in PFC2D. Itasca Consulting Group, Minneapolis
50. Zhou Y, Wu S, Xu X et al (2013) Particle flow analysis of acoustic emission characteristics in rock fracture process. Chin J Rock Mech Eng 32(5):951–959
51. Zhao T, Yin Y, Tan Y et al (2014) Microscopic simulation experiment study on coal and rock impact tendency based on particle flow theory. J Coal 39(2):280–285

Open Access This chapter is licensed under the terms of the Creative Commons Attribution-NonCommercial-NoDerivatives 4.0 International License (http://creativecommons.org/licenses/by-nc-nd/4.0/), which permits any noncommercial use, sharing, distribution and reproduction in any medium or format, as long as you give appropriate credit to the original author(s) and the source, provide a link to the Creative Commons license and indicate if you modified the licensed material. You do not have permission under this license to share adapted material derived from this chapter or parts of it.

The images or other third party material in this chapter are included in the chapter's Creative Commons license, unless indicated otherwise in a credit line to the material. If material is not included in the chapter's Creative Commons license and your intended use is not permitted by statutory regulation or exceeds the permitted use, you will need to obtain permission directly from the copyright holder.

Chapter 4
Key Technologies for Controlling Mining Dynamic Disasters

Deep coal resources are the main backup energy security for our country in the future. China is the country with the largest scale, difficulty, and number of deep coal mine construction in the world. The safe and efficient mining of deep coal resources is a key issue that most coal mines in our country must face. Typical mining dynamic disasters include rock burst, mine water inrush, and large deformation of roadway surrounding rock, which are sudden, sharp, and violent. They are extremely complex dynamic phenomena that occur in coal mine mining, essentially induced by stress field disturbance or strain increase during mining process, leading to the birth, development, and penetration of micro-fractures until instability occurs [1–5]. Mining is the root cause of dynamic disaster accidents.

As we enter deep mining, the severity of related dynamic disasters and mining depth show a non-linearly increasing development trend. We face new unfavorable mining environments such as high original rock stress and high mining-induced secondary stress. In addition, the non-linear characteristics of mining rock mechanical behavior become significant, and the sensitivity of the rock mass to mining disturbance and external dynamic response increases, making the mechanism of rock mass mining non-linear response more complex. The interaction between rock burst and coal and gas outburst disasters is highlighted, the inducing effect of mine tremors and strong earthquake stress fields on deep well coal mine dynamic disasters is enhanced, the accumulation and disaster formation of underground water bodies are difficult to detect in time, and the risk of large-scale roof pressure disasters increases. All types of dynamic disasters not only show a gradually rising trend in frequency, intensity, and scale, but the phenomenon of multiple disasters coupling and concurrent occurrence will also become more prominent, especially secondary disasters induced by dynamic disasters have occurred more frequently in recent years. The frequent occurrence of coal mine dynamic disaster accidents poses a serious challenge to the safe and green mining of coal in our country. So far, the prevention and control of dynamic disasters is still a worldwide problem, with great difficulty and high risk. Its control level is directly related to the overall level of coal mine safety development. Effectively controlling the occurrence of major dynamic disasters driven by engineering forces in deep mining fields is an important issue in the construction of our country's deep coal mine safety guarantee system.

In the past half-century, numerous domestic and international scholars and scientists have conducted extensive research on the mechanisms of dynamic disasters, forecasting and prediction technologies, and prevention and control technologies, laying a theoretical foundation and applied technology for guiding production practices, and making significant contributions to the safety of the mining industry. Overall, dynamic disaster accidents such as rock burst are closely related to mining depth, that is, the increase in overburden self-weight pressure and structural stress caused by the increase in mining depth, manifested as intense mine pressure in the mining area, severe deformation of the surrounding rock, instability of the roadway and mining area, and prone to destructive dynamic disaster accidents. Summarizing the reasons, no matter what kind of dynamic disaster, it is because mining activities cause changes in the stress state of the coal-rock mass, and under the action of mining stress, the coal-rock mass causes damage and destruction, which leads to instability and induces dynamic disasters [6]. Focusing on the current status of dynamic disaster research, this paper elaborates on the research of the disaster-causing mechanism and key prevention and control technologies of typical dynamic disasters in coal mines, in order to help fundamentally reverse the frequent occurrence of disaster accidents such as rock burst, roof water penetration, and large deformation of roadway surrounding rock in China.

4.1 Key Technologies for Rock Burst Disaster Prevention and Control

The first case of rock burst in China occurred in Fushun Shengli Coal Mine in 1933. Data shows that the number of rock burst mines and the provinces covered in China are constantly increasing. The essence of rock burst is the phenomenon of elastic energy release in a short time under high stress state of coal or rock. Depending on specific conditions, it manifests as a large amount of coal and rock powder being thrown out after crushing, overall displacement of the coal seam, brittle failure of coal and rock, instantaneous elastic vibration or loud noise of the coal seam or surrounding rock. These forms can appear alone or several forms can appear at the same time, thus causing different forms of rock burst manifestation.

4.1.1 Overview of Rock Burst Disasters

Describing the rapid rupture and instability of rock mass in mining and underground rock engineering, and the dynamic phenomena such as impact, vibration or sound, the names used in academic journals include rock burst, rock burst, mine earthquake, coal cannon, coal explosion, microseismic, rock burst, etc. The concepts are distinguished and interconnected. "Rock burst" is a mine pressure (dynamic) manifestation characterized by coal (rock) protrusion under mine pressure around the mining space, and is a major accident disaster in coal mines. In coal (rock) with "impact tendency" that stores high-strength compressive elastic energy, especially in areas where energy is concentrated, the excavation of tunnels and the advancement of mining faces (i.e., mining) trigger the release of corresponding elastic energy is the root cause of rock burst. The high-strength compressive elastic energy stored in the surrounding rock of mining, including the compressive elastic energy in coal (rock) and the bending (compression) elastic energy of the overlying rock (roof and floor) rock layer in the mining space, is the main force of rock burst. Therefore, first of all, the related concepts of rock burst are explained, a simple description of each concept is given, and the reasons and conditions for the occurrence of rock burst are analyzed in detail in combination with mine pressure theory and mine dynamics theory, summarizing the connection between mining stress and rock burst accidents, providing a theoretical basis for on-site analysis and handling of rock burst disaster accidents.

1. Research on Related Concepts and Classification Methods of Rock Burst

We often refer to the phenomena that occur in coal mine coal seam tunneling and coal mining faces as rock burst or mine burst, regardless of whether they occur in the coal body or in the rock layers of the roof and floor. The phenomena that occur in the deeper parts of the coal body but can be heard and felt slight vibrations are called coal outburst or coal cannon; those that occur in hard rock engineering and coal mine rock tunnel engineering are called rock burst; seismology classifies rock body ruptures observable by microseismic equipment as mining-induced earthquakes, simply referred to as mine earthquakes, but mining and engineering circles tend to call strong energy releases that occur far from the mining area mine earthquakes, to distinguish them from microseismic events that do not produce obvious impacts and vibrations. The new 2016 version of the "Coal Mine Safety Regulations" also treats the two terms rock burst and rock burst as equivalent, defining them as sudden, violent destructive dynamic phenomena produced by the instantaneous release of elastic deformation energy in the coal and rock bodies around the wellbore or working face, often accompanied by phenomena such as coal and rock body ejection, loud noise, and air waves. As understanding of the instability phenomena and mechanisms of rock body dynamic destruction deepens, most experts and scholars believe that although dynamic disasters such as rock burst, rock burst, and mine earthquakes have certain similarities, or can be used interchangeably to some extent, they have substantial differences in terms of phenomena, the lithology of the constituent media, and the mechanisms and control methods of occurrence.

They should be developed in conjunction with the engineering background, rather than avoiding the engineering background and talking about their mechanical nature being the same. Rock burst and rock burst should be used differently, paying attention to distinguishing the usage environment and scope of different terms, so as to make the research content more targeted [7, 8].

(1) Rock burst

Rock burst refers to the mining pressure (dynamic) manifestation characterized by coal (rock) protrusion around the mining space under mine pressure, which is a major accident disaster in coal mines. Due to the relative temporariness of mining engineering, whether this dynamic phenomenon has the "disaster destructiveness" that affects safe production is usually taken as the sign of rock burst occurrence.

Professor Qi [9] summarized the basic characteristics of rock burst phenomena based on the current situation of rock burst in China's coal mines: rock burst often occurs in the advanced tunnels during the recovery period, usually within 0–80 m of the advanced working face. After the occurrence of rock burst, a large area of the coal wall is peeled off, and coal is thrown out from the coal body. The occurrence of rock burst in coal mines usually occurs within the influence range of mining stress in front of the working face. Under the influence of stress concentration and mining, it leads to the occurrence of rock burst (Fig. 4.1). The coal and rock bodies where rock burst occurs, the roof and floor of the coal seam do not occur or significantly occur damage and deformation after the occurrence of rock burst, but the coal body does occur damage and moves out as a whole, and there are obvious sliding scratches and delamination (delamination height is about 0.1–0.15 m or even larger) between the coal seam and the roof and floor. Rock burst often occurs during the period of roof pressure, support movement or pillar release, blasting and other technological processes. Rock burst often occurs near geological structures such as thin coal seam belts, faults, folds, etc., where the structural stress is relatively large. The coal and rock layers where rock burst occurs have typical "three hard" structural characteristics, namely hard coal, hard roof and hard floor; and there is often a thin layer of powdery soft coal (about 0.1–0.2 m thick) between the roof and the coal seam. After the occurrence of rock burst, the cross-section of the tunnel shrinks significantly, usually reaching 50–70%, even up to 90%.

(2) Rock burst

Rock Burst (Rock Burst) is a dynamic instability geological disaster that occurs during the excavation process of underground engineering under high ground stress conditions, where hard brittle surrounding rock causes the elastic strain energy stored in the rock body to suddenly release due to excavation unloading, resulting in explosive loosening, spalling, ejection and even throwing [10, 11]. Dr. Gong Fengqiang suggests that in the specific study of rock burst, the "rock burst" mentioned in coal mines should be replaced with "coal burst"; while the "rock burst" in the general sense refers specifically to the rock burst in deep hard rock engineering.

Qian [12], based on the mechanisms and definitions of rock burst by five international authoritative scholars, divides rock burst into fault slip or fracture slip type

4.1 Key Technologies for Rock Burst Disaster Prevention and Control

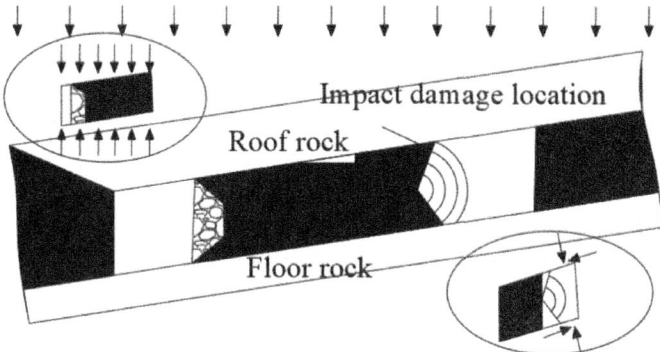

Fig. 4.1 Description of tunnel rock burst phenomenon

and strain type caused by rock failure, and believes that the mechanism of rock burst is the same as that of fracture slip type or shear type rock burst; Tang [12, 13] points out that the essence of rock burst is the dynamic process caused by the disturbance of stress wave after the quasi-static structure is close to the critical state, and rock burst occurs when the decrease of structural bearing capacity is higher than the decrease of structural stiffness; He [14] analyzes from the energy perspective that the rock failure process includes failure energy and surplus energy, the failure energy is mainly used for rock failure and crack formation, the surplus energy induces rock burst, and further divides rock burst into excavation type and excavation-impact induced type.

Qi [15] summarized the rock burst phenomena in metal mines and hydropower tunnel projects, and concluded that the rock burst phenomena have the following typical characteristics: when rock burst occurs, it is usually accompanied by obvious sound characteristics, and the sound varies with the scale of the rock burst. The rock damage caused by rock burst is characterized by plate-like ejection. Usually, the ejection distance of rock caused by rock burst is mostly within 5 m, and the ejection distance of larger rock burst can reach more than 10 m. The occurrence of rock burst is closely related to the direction of ground stress (Fig. 4.2).

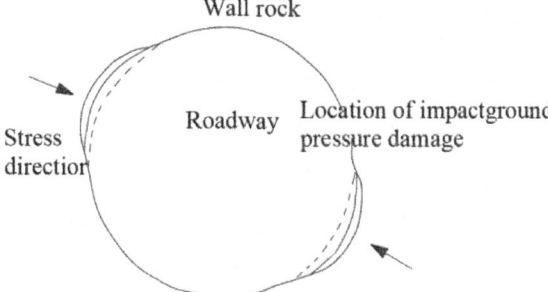

Fig. 4.2 Description of rock burst phenomenon

(3) Mine tremor

Literally [9], mine tremor mainly refers to the vibration events that occur during the mining process, especially refers to the earthquakes induced by mining or the stratum vibration that occurs under the conditions of mine engineering. In terms of subordination, the deformation of the roadway and other dynamic phenomena accompany the occurrence of rock burst, and there will be stratum vibration in this process, so mine tremor always accompanies the occurrence of rock burst, but the converse is not true. The main feature of mine tremor is the vibration events that occur in mine engineering, while the main feature of rock burst is the destructiveness and disaster of the mining space. It cannot be simply considered that mine tremor and rock burst are equivalent. At the same time, the environment of mine tremor includes coal mines, gypsum mines, metal mines and other mining environments, while rock burst is generally used in underground coal mining environments, and their applicable ranges are also different.

(4) Coal bump

Coal bump refers to the sound (muffled thunder, machine gun sound, rustling sound, etc.) produced by the misalignment of the deep part of the coal body during the coal seam mining process due to rock vibration, roof fracture or small-scale rock deformation unloading coal seam mining pressure [16], often as a sound precursor to coal and gas outburst. The sound of coal bump usually appears in the stress concentration area, such as geological structure belt, coal pillar influence area or mining stress concentration area. When the sound of coal bump comes from far to near, it indicates that coal and gas outburst may occur immediately.

(5) Microseismic

Microseismic mainly refers to the small vibrations caused by rock fracture or fluid disturbance. In a broad sense, microseismic can be divided into engineering microseismic (Micro Seism) and natural microearthquake (Micro Earthquake). The microseismic phenomenon that appears in the process of rock burst in coal mines is mainly engineering microseismic, which is characterized by low frequency (3–30 Hz), large energy, slow attenuation, and long propagation. In essence, microseismic waves can be regarded as earthquake waves with less energy; in terms of frequency research range, from small to large, they are earthquake, seismic prospecting, microseismic, and acoustic emission, but the research ranges of these four are overlapping, not clearly defined. Since the process of rock burst is accompanied by the fracture of coal and rock, microseismic signals will be generated, so microseismic monitoring technology can be used for source location, crack identification, impact risk warning, etc. [17, 18]. Rock burst always accompanies microseismic events, but microseismic events do not necessarily lead to rock burst.

The mechanical environment, location, macro and micro manifestations of rock burst are diverse, and the intensity and degree of damage caused by rock burst are also different. Rock burst can be classified according to stress state, manifestation intensity, and different locations and positions. The main classification methods of rock burst are as follows.

4.1 Key Technologies for Rock Burst Disaster Prevention and Control

Classification by energy characteristics of rock burst. According to the seismic energy released during the burst, rock burst is defined as microburst, weak burst, medium burst, strong burst, and disastrous burst. Microburst mainly manifests as small-scale rock ejection and mine microseismic, including fall and microseismic. Fall is a local surface damage, manifested as a single coal rock block popping out, accompanied by a shooting sound. Microseismic is a local damage in the deep part of the coal body that does not cause crushing and ejection, often accompanied by sound and rock microseismic. Weak burst often manifests as local damage with a small amount of coal rock ejection, accompanied by obvious sound and seismic effects, but does not cause serious damage. The surrounding rock produces a vibration of less than 2.2 magnitude. Medium burst refers to a sudden brittle failure, throwing out a large amount of rock, forming an air wave, causing damage and collapse of several meters long roadway support, and moving or damaging mechanical and electrical equipment. A strong burst can cause a part of coal or rock to break sharply, a large amount is thrown into the mined space, there is support fracture, equipment movement and surrounding rock vibration, the magnitude is above 2.3, accompanied by a huge sound, forming a large amount of coal dust and generating a shock wave, which can cause the collapse of the roadway support that is several tens of meters long, damage to mechanical and electrical equipment, and requires a large amount of repair work. A disastrous burst can cause the entire mining area or a tunnel within a level to collapse, and even in some cases affect the entire mine, causing the entire mine to be scrapped.

According to the type of rock involved in the burst, rock burst can be divided into coal layer burst and rock layer burst. According to the location of the rock burst, it can be divided into coal layer rock burst, roof rock burst, and floor rock burst. Occurring in the coal body, according to the depth and intensity of the burst, it is further divided into surface, shallow, and deep bursts. Classification by workspace. According to the workspace, rock burst can be divided into rock burst occurring in the excavation face and rock burst occurring in the coal mining face.

According to the source of the burst force, rock burst is divided into gravity type, structural type, and intermediate type. Gravity stress type rock burst: mainly caused by gravity, with no or only minimal structural stress influence. Structural stress type rock burst: mainly caused by structural stress (structural stress far exceeds the self-weight stress of the rock layer). Intermediate or gravity-structural type rock burst: mainly caused by the combined action of gravity and structural stress.

According to the magnitude intensity and considering the amount of ore thrown out, rock burst can be divided into three levels: slight burst (Level I), indicating that the amount of ore thrown out is less than 10t, and the magnitude is less than 1. Medium burst (Level II), indicating that the amount of ore thrown out is between 10–50t, and the magnitude is between 1–2. Strong burst (Level III) indicates that the amount of ore thrown out is more than 50t, and the magnitude is more than 2.

In addition, according to the source of stress and loading form of rock burst, that is, the classification method of starting conditions highlights the role of force source factors on rock burst, which is most relevant to the mechanism and prevention research of rock burst. According to this classification basis, Dou and He [19] divided rock burst into mining type rock burst caused by mining activities and structural type rock burst caused by structural activities, and mining type rock burst can be divided into pressure type, impact type and impact pressure type, structural type rock burst can be mainly divided into fold type and fault type; Pan [20] and others re-divided from the starting conditions of the burst, dividing rock burst into concentrated static load type and concentrated dynamic load type.

2. Causes and conditions of rock burst accidents

Theory and practice have proven that not all coal seams will experience rock burst during mining. Macroscopically, three conditions need to be met for a rock burst to occur during the mining process. Condition one: The coal seam being mined has a tendency for rock burst, which reflects the inherent ability of the material to cause shock-type damage. The tendency for rock burst in coal seams can be measured by the degree of rock burst tendency. The national standard GB/T25217.2-2010 stipulates the classification and determination methods of coal's rock burst tendency index; Condition two: The load of the surrounding rock in the mining space where rock burst occurs is localized and concentrated, and it reaches the shock-type instability load limit of the coal rock structure system. There are two situations where the load is highly localized and concentrated. One is that it is a high-stress concentration area itself, and rock burst occurs as soon as the roadway is excavated to create a release space. The other is due to the excavation of the roadway and the working face, which causes the stress to be redistributed and the load to be superimposed, localized, and concentrated, and rock burst occurs; Condition three: The highly localized and concentrated load in the coal body has a release space. The process of damage to the coal rock body, equipment, and personnel in the mining space is the process of releasing the highly concentrated load. In this case, if no mining is carried out and no voids or other release spaces are formed, rock burst will not occur.

It is generally believed that the main influencing factors of rock burst include the conditions of the top and bottom plates of the coal seam, the original rock stress, the depth of the seam, the physical and mechanical properties of the coal seam, the thickness and inclination angle, etc. Although there are records of rock bursts occurring in extremely shallow hard coal seams, current statistics still show that as the mining depth increases, the frequency and intensity of rock bursts also increase. Pan et al. [21] believe that the occurrence of concentrated static load type rock burst is mainly characterized by slow stress migration, concentration, and progressive loading. The main influencing factors include: the increase in self-weight stress due to the increase in mining depth; the increase in horizontal structural stress due to historical tectonic movements; the superposition of mining stress due to adjacent or opposite mining, isolated coal pillars; the concentration of mining stress in advance or the concentration of mining stress on the side of the roadway; the change in coal seam thickness leading to local thinning or sharpness causing stress concentration; faults causing the concentration of stress in the upper and lower plates of the fracture zone; the speed of mining or excavation is too fast, causing the coal rock body stress to be unable to adjust, etc.; the occurrence of concentrated dynamic load type rock burst is mainly characterized by pulse load or elastic wave loading, and its main

4.1 Key Technologies for Rock Burst Disaster Prevention and Control

influencing factors include: large-area roof breaking and sliding in the mined-out area of the working face; roof breaking caused by a large amount of coal pillar recovery; "activation" of faults near the working face; vibration waves generated by underground blasting; disturbances caused by natural earthquakes.

The high-strength coal (rock) layer, due to the high degree of stress concentration and the storage of high-energy-level elastic deformation energy formed by tectonic movement and mining field advancement, is the fundamental cause of rock burst. Without taking measures to release stress and energy, advancing the mining face in places where there may be high stress concentration and high energy level elastic energy release is the condition for the realization of rock burst.

The high-energy-level elastic energy stored in high-strength coal (rock) with a tendency to rock burst includes the elastic energy stored by coal (rock) being squeezed by tectonic movement, the compressive elastic energy gathered under the condition of a hard roof on a large area of the mining field, and the bending deformation elastic energy of a large area of high-strength, thick, hard rock layer. Therefore, understanding the history of coalfield tectonic movement and the real distribution of residual tectonic stress, mastering different mining methods, different mining parameters, and different mining procedures under specific coal seam conditions, and the impact on coal (rock) stress and energy accumulation and release, is the key to preventing rock burst. Its stress and energy criteria are:

Stress criterion:

$$\sigma \geq \sigma_C$$

In the formula, σ is the stress of the coal body; σ_C is the uniaxial compressive strength.

The elastic energy can be calculated by the following two formulas for the unit volume of elastic energy formed by the volume change and shape change of the coal seam:

$$W_V = \frac{(1-2\mu)(1+\mu)^2}{6E(1-\mu^2)}(\gamma H)^2 \tag{4.1}$$

$$W_\phi = \frac{(1-2\mu)^2}{6G(1-\mu^2)}(\gamma H)^2 \tag{4.2}$$

If only considering the effect of gravity, then the total energy W of the unit coal body at depth H is the sum of W_V and W_ϕ:

$$W = \frac{(1+\mu)(1+2\mu)}{2E(1-\mu)}(\gamma H)^2 \tag{4.3}$$

Considering the stress concentration coefficient of the support pressure zone, the above formula can be written as:

$$W = \frac{(1+\mu)(1-2\mu)}{2E(1-\mu)}(K\gamma H)^2 \tag{4.4}$$

The energy of the broken unit volume U_2 is:

$$U_2 = k_0 \frac{\sigma_C^2}{2E}. \quad (k_0 > 1) \tag{4.5}$$

Under the condition of impact ground pressure energy, we have:

$$W > U_2 \tag{4.6}$$

3. The relationship between mining stress and impact ground pressure accidents

The mining stress field is closely related to the original rock stress, and it is also affected by various factors such as rock body structure, properties, and mining space size. Its distribution in space has a certain range and continues to evolve with the progress of mining activities and the passage of time.

The study of the mechanism of impact ground pressure is the basis for its effective prevention and control. Domestic and foreign experts and scholars have conducted in-depth research based on elasticity, plasticity theory, and stability theory, and have proposed stiffness theory, strength theory, energy theory, impact tendency theory, deformation system instability theory, shear slip theory, three criterion theory, "three factors" theory, strength weakening and damping theory, composite thick coal seam "shock" mechanism, rock mass dynamic instability folding mutation mechanism, impact initiation theory, coal-rock combination impact mechanism, impact ground pressure and outburst unified instability theory, etc. These theories reveal the conditions and principles of impact ground pressure from different aspects, promoting the study of impact ground pressure.

The occurrence of impact ground pressure is a result induced by multiple factors. Regardless of the theory, it ultimately cannot be separated from the most relevant mining stress conditions for impact ground pressure; at the same time, the three main factors leading to the occurrence of impact ground pressure: internal factors (impact tendency of coal and rock), force source factors (high stress concentration or high deformation energy storage and external dynamic disturbance) and structural factors (weak structural surface and layer interface prone to sudden sliding) are also closely related to mining stress. Therefore, the study of mining stress and its loading form is of great significance to the theory and prevention research of impact ground pressure.

(1) The relationship between the occurrence of impact ground pressure in excavation and mining stress

The occurrence of impact ground pressure in excavation occurs during the advancement of the excavation working face. Among them, the conditions for the occurrence of impact in coal roadway excavation in the original stress field are: ① the coal seam strength is high ($f > 1.5$), the water content is low (<3%), and brittle failure occurs

4.1 Key Technologies for Rock Burst Disaster Prevention and Control

under pressure, that is, "there is a tendency for impact failure"; the coal seam thickness is large, generally exceeding 2 m; the stress in the roadway surrounding rock (including coal seam and roof and floor) reaches the limit of impact failure. The "critical mining depth" to reach this limit under the condition of a single gravity stress field is generally above 700–800 m. For the original stress field with structural stress, when the mining depth exceeds 500–600 m in the thick coal seam, there may be accidents of roof coal impact failure.

In the stress field affected by mining, the conditions for the occurrence of impact ground pressure in the coal seam and the stress limit requirements are the same as those for the roadway excavation in the original stress field. However, when considering the conditions for the realization of the limit stress, it can no longer be simply linked with the size of the original stress and the corresponding mining depth. On the contrary, it is necessary to put the understanding of the characteristics of the redistributed stress field at different mining depths and under different mining conditions, and its formation and development rules in the first place. Compared with the excavation of roadways in the original stress field, when judging the stress conditions required for the possible realization of impact ground pressure in the excavation of roadways in the mining stress field, it is necessary to grasp the following major differences in the mining stress field under different mining conditions.

After the mining area advances, the stress in the original gravity stress field will be redistributed. The redistributed stress field, according to the difference in stress size, is divided into "low stress area" ($\sigma < \gamma H$), including the "internal stress field" directly connected by the gravity of the fractured moving rock layer and a part of the stress in the plastic failure zone is less than the original stress. The "high stress area" includes parts where the stress in the elastic–plastic zone exceeds the original stress, and the "original stress field" that is not affected by mining. Obviously, under the same mining depth conditions, in the mining stress field, the size of the surrounding rock stress and the possibility of impact ground pressure in the excavation of roadways in different locations will have major differences. If a roadway is excavated in a high-stress area, the "critical mining depth" for the occurrence of impact ground pressure may be reduced by half compared to the excavation of a roadway in the original stress field. On the contrary, there is no possibility of impact ground pressure in the roadway excavated in the stable internal stress field, regardless of the mining depth.

After the mining area advances, the structural stress in the original stress field will be released to varying degrees depending on the changes in mining conditions and the development of corresponding rock layer movement and destruction. For example, in the case of extra-thick coal seams and shallow coal seam burial, the "reverse procedure" mining from bottom to top is adopted, and under normal mining procedure conditions, the realization of excavation in the stable "internal stress field" does not need to consider the danger of structural stress causing impact ground pressure.

Under the same mining depth conditions, the size and distribution characteristics of the mining stress field are controlled by the length of the working face and the mining height. Within a limited range of working face length, the range of mining stress distribution and the range of each corresponding different stress interval will increase with the increase of working face length and mining height. Therefore, when choosing the location of tunnel excavation to avoid the occurrence of rock burst, attention must be paid to the significant differences brought about by changes in the length of the working face and mining height conditions.

The formation and stability of the mining stress field have a development process related to mining conditions, which is directly related to the development and stability of the corresponding rock layer movement. Therefore, the achievement of the goal of avoiding stress overload in the excavation of tunnels in the mining stress field not only depends on the location of tunnel excavation, but also requires great attention to the timing of tunnel excavation. The location and timing of tunnel excavation should be dialectically unified. Compared with excavating tunnels in the original stress field, only by mastering the significant differences in the mining stress field under different mining conditions can we better judge the possible stress conditions required for rock burst in the excavation of tunnels in the mining stress field.

The rock burst induced by excavation, and the vibrational impact damage, not only occur at the source of excavation—near the working face, but also affect all tunnels and workplaces in the entire high-energy compression (high stress concentration) area. The larger the range of stored compressive elastic energy, the higher the energy level, the greater the intensity of impact damage and the range of impact. Therefore, on the basis of clarifying the size, distribution and formation development rules of stress in the related stress field, the maximum avoidance of excavation and tunnel layout in the area of high pressure energy accumulation is the key to controlling rock burst damage and corresponding accidents. Among them, for the original stress field with structural stress, without taking measures to release structural stress, it is absolutely necessary to avoid excavation and tunnel layout on the structural axis with high energy compression stress. In the design of mining schemes in the structural stress field, including mining area division, uphill mining area and cutting eye layout, as well as the selection of working face advancement direction and length, the following principles should be followed to achieve the above goals.

Advance the mining face perpendicular to the structural axis, ensure all mining tunnels are arranged perpendicular to the structural line, and pass through the parts storing compressive stress in the shortest time and length; try to expand the size and length of the mining area and working face advancement, and maximize the exclusion of the possibility of excavating a large number of uphill mining areas and cutting eyes in the compressive stress area near the structural axis; strive to arrange the cutting eye, mining flat tunnel, uphill mining area and other tunnels in the internal stress field where the compressive stress has been basically released after the advancement of the mining face, to achieve post-mining excavation.

In the mining stress field with a single gravity, it is absolutely necessary to avoid excavating tunnels in the coal pillars storing high compressive energy and the peak area of mining stress in the mining area. It is necessary to strive to arrange the uphill mining area, cutting eye and mining flat tunnel in the "internal stress field" that has stabilized after the advancement of the mining face, and to excavate tunnels in the stable internal stress field. Figure 4.3 shows three possible locations for the excavation of mining flat tunnels and cutting eyes under the condition of three-sided

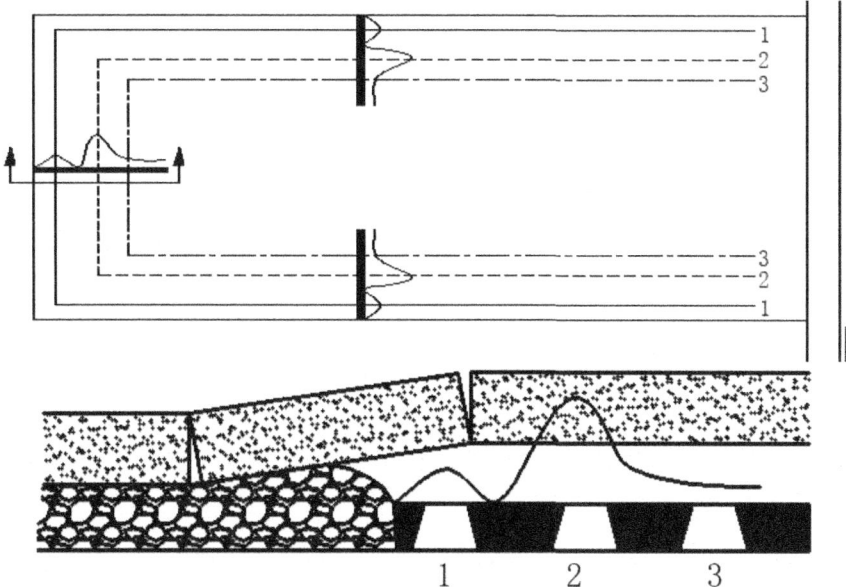

Fig. 4.3 Diagram of tunnel location in mining stress field

mining empty coal seams. Among them, 2 is the excavation of tunnels in the stress peak area, rock burst is inevitable; 3 is the excavation in the original stress field, rock burst will only occur when the mining depth reaches the critical value; 1 is the excavation in the internal stress field, there will be no rock burst accident.

(2) The relationship between the occurrence of rock burst during mining and mining stress

Rock burst during mining occurs during the advancement of the mining face. The conditions for the occurrence of rock burst during mining are: the coal seam has a "tendency to impact damage", and the stress and elasticity accumulated in the coal seam and overlying rock layers can reach the limit sufficient to produce impact vibration and surrounding rock damage.

The force (energy) source of rock burst on the mining face includes the compressive elastic energy of the coal seam under the pressure of the large-area exposed overlying rock layer, and the elastic energy stored by the elastic bending deformation of the high-strength large-thickness hard roof. The deeper the burial depth of the coal seam and the larger the exposed area determined by the high-strength hard roof, the higher the corresponding energy level.

The higher the energy induced by mining and the energy induced and released by the self-fracture of the roof, the larger the range of impact, and the more serious the impact damage in the working face and adjacent tunnels. The rock burst that occurs during the advancement of the mining face includes two types: the release of compressive elastic energy in the coal seam induced by mining or basic roof fracture, and the rock burst caused by the fracture of the thick coal seam hard roof in a large area of elastic bending state.

The impact ground pressure caused by the release of compression energy in the coal seam in the backfilling working face occurs at the high stress concentration areas around the mining area. The closer the working face and adjacent roadway are to the epicenter, the greater the risk of vibrational impact damage and corresponding accidents. Conversely, the further the working face and adjacent roadway are from the epicenter, the wider the "buffer zone" for absorbing (reducing) elastic energy, the smaller the threat of vibrational impact damage and related accidents. Therefore, it is very effective to arrange the backfilling roadway in the "internal stress field" as much as possible, and to take measures such as pre-water injection or blasting loosening to increase the buffer zone width at the part of the working face that may experience impact ground pressure, reducing the threat of impact ground pressure, especially "vibrational impact damage". This is also the theoretical basis for controlling impact ground pressure damage accidents effectively by maximizing the mining height and especially using top-coal caving technology when mining thick coal seams with impact ground pressure danger. The range of impact ground pressure damage induced by the advance of the backfilling working face and its relationship with the stress distribution of the mining area are shown in Fig. 4.4.

In Fig. 4.4, S_0 represents the low-stress interval that can buffer the compressed and damaged coal seam. When impact ground pressure occurs, the coal wall and

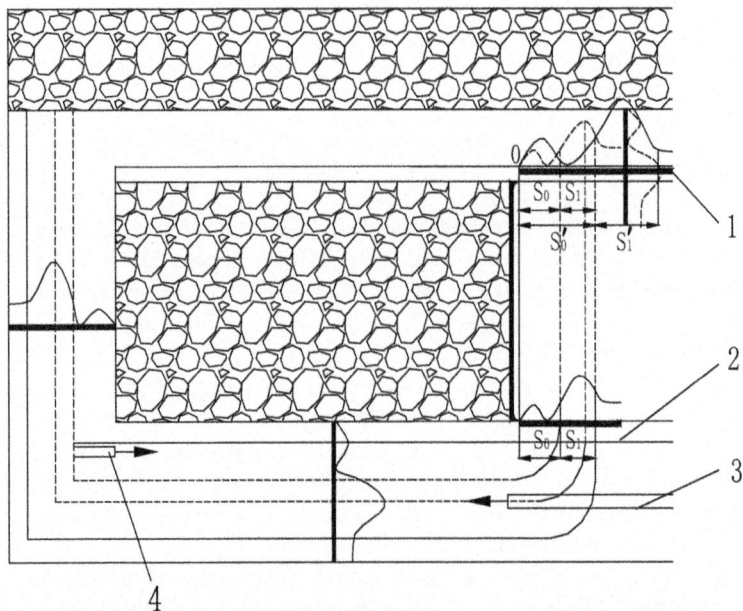

Fig. 4.4 Diagram of impact ground pressure and mining stress distribution induced by the advance of the backfilling working face

roof and floor of the working face and the advanced roadway in this interval (1 and 2 in the figure) will only experience mild movement damage, posing little threat to personnel and equipment safety. S_1 represents the high-energy elastic compression stress area, which is the source of impact ground pressure. When impact ground pressure occurs, the roof and floor and both sides of the roadway in this range will experience strong impact damage, which can easily cause casualties. The size of the impact damage area in the advanced flat roadway (S_1) depends on the mining stress value determined by the mining depth and the exposed area of the old pond rock layer. Among them, the advanced flat roadway near the upper mined-out area (1 in Fig. 4.4), if it is located at the peak of the mining stress of the upper working face, will experience a range and intensity of impact damage that is multiples greater than the lower advanced roadway due to the superposition of the compression stress of the two working faces. Conversely, if the roadway is located in the low-stress area of the mining stress distribution of the upper working face, i.e., the "internal stress field", its impact damage intensity and range (S_1) will be much smaller than the lower advanced flat roadway (2 in Fig. 4.4). Similarly, when excavating and deploying the backfilling flat roadway of the lower continuous working face, if the roadway is mistakenly placed in the peak area of the mining stress distribution of this working face and excavated in advance (as in 3 in Fig. 4.4), it is very likely to face the risk of impact ground pressure accidents induced by excavation when the excavation working face enters the high-stress area of mining stress, and all roadways within the full length range of the upper working face's advance will have to withstand the threat of impact ground pressure damage from the advance of the working face and throughout the entire process of the continuation working face's production advance, the full length range of the roadway will be threatened by backfilling impact ground pressure. Therefore, the correct layout scheme for the backfilling flat roadway of the continuation working face should be to place it at position 4 in Fig. 4.4, achieving excavation in a stable internal stress field.

Under the conditions of medium-thick coal seams, the type of damage to the working face and roadway surrounding rock caused by the release of compressive elasticity is mainly the squeezing and spouting of the coal side. Under the conditions of layered mining of thick coal seams, when the impact ground pressure of the top mining layer occurs, there will be serious cases of large-scale bottom coal vibration damage in the working face and advanced roadway.

The large-area hard roof bending elastic energy release type impact ground pressure of the backfilling working face advance occurs at the moment of thick hard roof fracture, and the strong impact damage includes the following two parts.

The first is the destruction of the working face and advanced roadway surrounding rock caused by the simultaneous release of coal seam compression elasticity at the peak of induced mining stress. The characteristics of the damage are the same as those of the single coal seam compression elasticity release, only the intensity is higher and the impact and damage range is larger. The closer the roof storing bending elastic energy is to the coal seam, the higher the corresponding impact damage intensity will be, and the range will also be larger. Secondly, the damage caused by the high-speed settlement of the roof at the moment of fracture, impacting and squeezing the coal wall. This kind of impact damage occurs at the working face and both ends of the roadway. The closer the energy-releasing roof is to the coal seam, the higher the damage intensity. And regardless of whether the coal wall of the working face has been damaged under the action of mining stress and how much the damage depth is.

The occurrence of rock burst that threatens safety production during the advancement of the mining face is conditional and regular. Practice has proven that the conditions for the occurrence of rock burst that threatens safety production are: the mined coal seam has "rock burst tendency"; the coal seam in the advancing part of the working face accumulates enough to produce rock burst destruction. The source of this compressive elasticity can be residual structural stress, or it can be peak stress formed by mining; during the coal breaking and roof falling production process, the roof breaking and pressing and other induced rock burst energy, reaching the limit that promotes the release of the elasticity of this part; the coal gangue of the working face and the two lanes ahead have not formed a sufficient buffer destruction interval. The requirement of having the above conditions at the same time determines the regularity of the time and place of destructive rock burst during the advancement of the mining face. Among them:

(1) The law of rock burst occurrence during the initial mining stage of the working face:

When the working face is arranged in the original structural stress accumulation position, or arranged in the peak area of the mining stress of the adjacent mined working face, in this case, from the beginning of the working face advancement, to the position where the compressive elasticity accumulated in this part is basically released (generally 8–10 m), there is a danger of destructive rock burst. Therefore, avoiding the arrangement of the cutting eye in the structural stress field or near the high-pressure area of the working face, or the advancement plan of arranging the cutting eye in the internal stress field formed by the mined working face, is the key to eliminating the destructive rock burst in the initial mining stage of the working face. Under the condition of existing structural stress, arranging the cutting eye in the internal stress field formed by the adjacent mined working face not only achieves the goal of safe excavation by releasing the structural stress in advance, but also eliminates the accident in the initial mining stage of the mining face.

In a single gravity stress field, the plan of arranging the cutting eye in the "internal stress" field formed by the mined working face and advancing the working face maintains the condition that the original advancing working face has enough buffer band width to avoid destructive rock burst. Avoid the danger of destructive rock burst in high stress area excavation and the beginning of working face advancement.

Arranging the cutting eye in the original stress field, the rock burst during the initial mining stage occurs at the position where the mining stress increases with the advancement of the mining field reaches the limit of rock burst destruction. The first fracture of the basic roof is the most important inducing force. For the rock burst of the release type of hard roof bending elasticity. The first fracture of the hard roof is the direct driving force for the occurrence of strong destructive rock burst.

4.1 Key Technologies for Rock Burst Disaster Prevention and Control 141

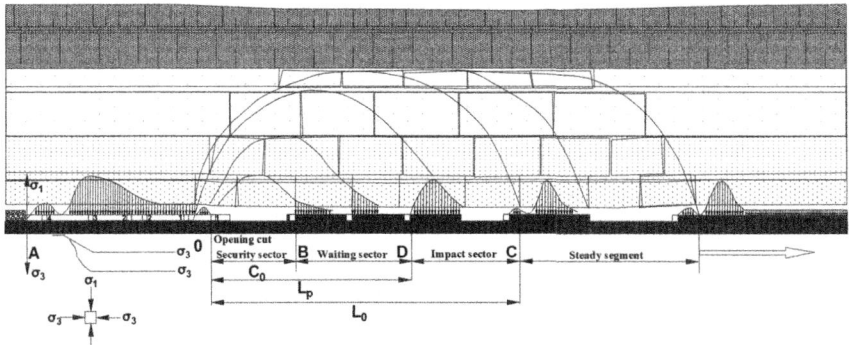

Fig. 4.5 Diagram of the law of rock burst occurrence during the normal advancement stage

(2) The law of rock burst occurrence and development during the normal advancement stage:

After the mining field enters the normal advancement stage, in terms of the possibility of rock burst, it includes the following two intervals: "Dangerous interval": It includes from the increase of concentrated stress on the coal wall to the beginning of rock burst destruction, to the coal wall destruction. The part where the advanced destruction has penetrated into the formation of a sufficient buffer band width. Within this section, both the high-energy-level mining-induced and basic roof fractures have the danger of causing destructive rock burst.

"Stable interval": that is, after the buffer zone is formed in front of the coal wall, the entire length until the completion of the working face advancement. In this interval, unless the hard roof breaks and releases the high-strength bending elasticity shock, the interior of the working face will not be destructively shocked. In the original stress field without structural stress, arrange the cutting eye, during the advancement of the working face, the law of the difference in the time and place of rock burst, as shown in Fig. 4.5.

4.1.2 Rock Burst Occurrence Mechanism

The essence of the rock burst phenomenon is the sudden instability and destruction of the coal and rock structure under high stress conditions, but due to the complexity of the mining space and mining technology, the current research on the mechanism and prevention technology of rock burst is not sufficient, and it is difficult to use one mechanism to explain the cause of rock burst, and it still needs to carry out long-term hard exploration and practice to possibly make breakthrough progress.

The academic community in our country analyzes the process of rock burst as a dynamic stability problem, and the research on the mechanism of rock burst can be roughly divided into three categories: the first category is to start from the research on the physical and mechanical properties of coal and rock materials, analyze the instability and destruction characteristics of coal and rock bodies and the inherent factors that induce their instability, and use chaos, bifurcation and other nonlinear theories to study the process of rock burst instability; the second category is to start from the research on the geological structure and deformation localization of the outburst area, analyze the relationship between the geological weak surface and the geometric structure of the coal and rock body and rock burst; the third category is to study the relationship between engineering disturbance (such as the vibration wave generated by blasting, etc.) and mining influence and rock burst. From the essence of the sudden instability and destruction of the coal and rock body caused by the stress state, the classification research on rock burst is carried out, Jiang [13] Coal mine shock pressure is divided into 3 types: material instability type shock pressure, slip fault type shock pressure, and structural instability type shock pressure. Material instability type shock pressure refers to the rock mass around the tunnel or working face during the excavation process, after the stress concentration in the coal rock mass reaches a certain level, the internal cracks in the coal rock material continuously expand, penetrate, and converge, causing a certain range of coal rock mass to undergo projectile, explosive destruction and resulting in shock outburst. The material instability type shock pressure is shown in Fig. 4.6a. Slip fault type shock pressure, as shown in Fig. 4.6b, refers to the coal seam slip fault outburst caused by the difference in stiffness between the roof and floor and the coal seam under the influence of mining, such as the coal seam planar outburst model studied by Lippmann [22]; or the sudden and violent destruction caused by the slip fault of faults, structures or structural surfaces near the tunnel. Structural instability type shock pressure refers to the sudden and violent destruction caused by mining stress or sudden rupture of a large area of the roof or induced by mine earthquakes, often resulting in a large area of coal pillar or tunnel surrounding rock outburst and the overall tunnel structure instability, as shown in Fig. 4.6c.

Jiang [23] Based on the difference between the main and guest objects, shock pressure is divided into "spontaneous type" and "induced type". "Spontaneous type" shock pressure refers to the shock damage that occurs when the stress around the mining space accumulates and meets the conditions of shock mechanics. For this type of shock pressure, the force source is "stress", and the shock pressure is the main body, which is actively occurring. "Induced type" shock pressure refers to the shock damage caused by the sudden accumulation of stress induced by far-field vibration when it meets the conditions of shock mechanics. Its force source is "vibration", such as structural activation, thick hard rock layer fracture, coal pillar destruction, blasting, and vibration caused by earthquakes. For this type of shock pressure, shock pressure is the object, it is induced by vibration. The study clarifies the process of shock pressure occurrence, points out the role of "stress" in the process of shock pressure occurrence, and points out the "control object" for shock pressure control; highlights the "force source" of shock pressure occurrence, providing a basis for studying shock pressure monitoring and early warning methods; points out the main body and mode of destruction of the two types of shock pressure, providing a basis for building a shock pressure monitoring technology system.

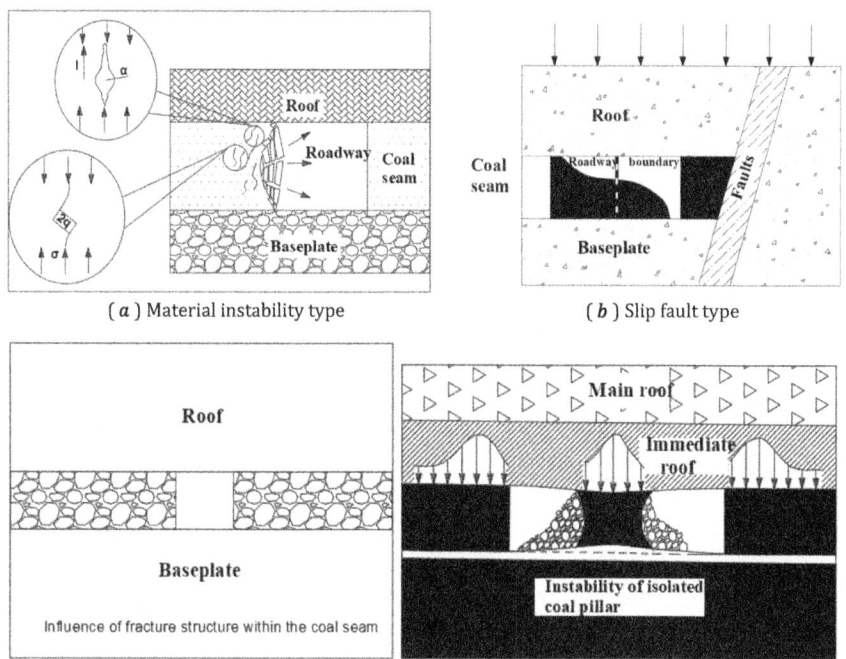

Fig. 4.6 Schematic diagram of shock pressure

Professor Pan [24] summarized the process of shock pressure occurrence into 3 stages, namely the shock initiation stage, the shock energy transmission stage, and the shock pressure manifestation stage, as shown in Fig. 4.7. The coal rock fracture zone that plays a leading role in the shock pressure manifestation stage is defined as the shock pressure initiation area, which corresponds to the first stage in the process of shock pressure occurrence, that is, the shock initiation stage. A high-energy event is detected by microseismic monitoring, followed by low-energy events, or the detected energy is particularly large, some reaching 10^8 J or more, but in fact, after the underground shock pressure, the shock pressure manifestation is not strong, and it does not cause much damage; sometimes, the located energy is small, but the shock pressure manifestation is very strong, therefore, it shows that there must be an energy transmission process from shock initiation to shock pressure manifestation stage, the transmission process may cause energy attenuation, this is the second stage in the process of shock pressure occurrence, that is, the shock energy transmission stage. From the perspective of preventing and studying the conditions of shock pressure occurrence, it is best to contain the shock pressure in the initial stage, that is, the shock initiation stage.

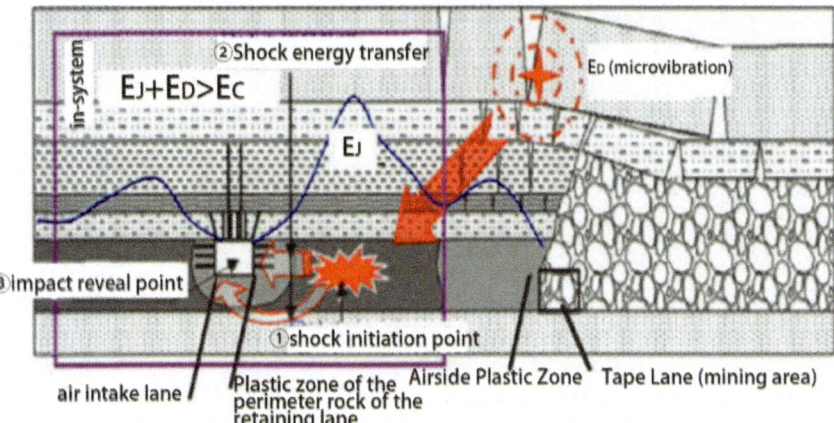

Fig. 4.7 Kinematic model of impact initiation along the empty lane [20]

The evolution process of contrast impact ground pressure is shown in Fig. 4.8. For the coal and rock mass with impact risk, the OA stage is when the internal fissures of the material are compacted; the AB stage is when stress and strain show an approximate linear growth, accompanied by volume changes, which is also the accumulation stage of the basic static load of the impact load body; the BC stage is the load aggregation stage, where the coal and rock mass stress and strain show a clear nonlinear growth, and the material's micro-cracks are constantly occurring and developing. There are two loading paths from this stage to the impact initiation point C, one is to continue to gain static load increments, and the other is to gain external dynamic load increments. The entire OC stage is the gestation stage before the impact starts, that is, the dynamic and static load loading stage, where there is a situation of premature death of the coal cannon due to insufficient load. Point C is the critical point, which is also the impact initiation point. At this time, $E_J + E_D - E_C > 0$ (E_J, E_D, E_C represent the concentrated static load, concentrated dynamic load, and the minimum load required for rock mass dynamic failure respectively). The CD stage is the stage where the remaining impact energy does work on the surrounding rock, which includes the destruction of the carrier and the barrier, that is, the manifestation of impact ground pressure.

Before mining, the coal and rock mass is in a deep three-dimensional stress balance state. Mining activities break the original stress balance, leading to the redistribution of the macro stress field and energy field in the three-dimensional space of the mining area. This dynamic evolution and development of the stress field and energy field inevitably create conditions for the gestation, occurrence, and development of dynamic disasters. Therefore, by studying the spatiotemporal evolution rules and multi-factor coupling disaster-causing mechanism of mining stress distribution and energy field, we can reveal the energy accumulation and release mechanism, spatiotemporal evolution rules of the energy field, and energy triggering conditions of deep fracture coal and rock mass during the mining process, and propose a model

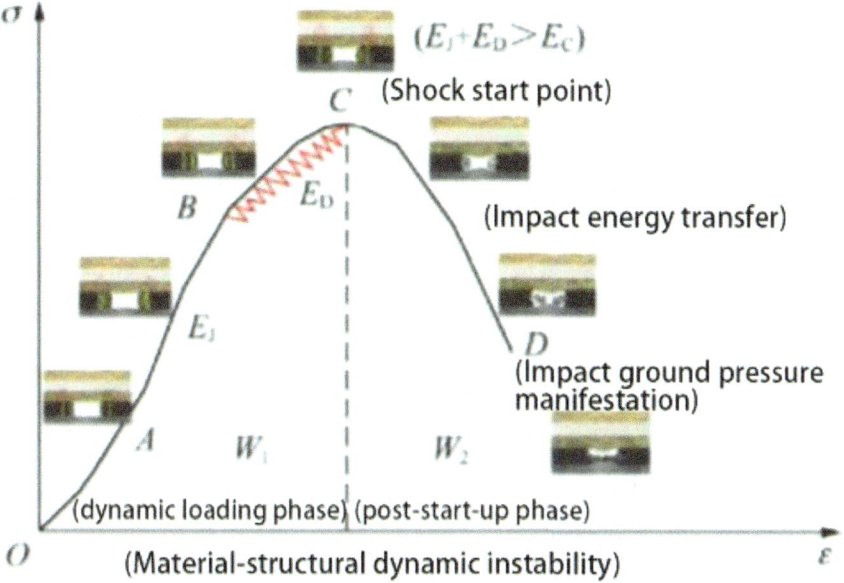

Fig. 4.8 Impact ground pressure and full stress–strain curve of rock mass [20]

and discrimination criteria for deep coal and rock mass dynamic instability based on energy mutation and an energy analysis system.

4.1.3 Basic Dynamic Information for the Prevention and Control of Impact Ground Pressure Disaster Accidents

1. Key technology for the prevention and control of impact ground pressure disasters

As mentioned earlier, there are three conditions for the macro occurrence of impact ground pressure during coal seam mining. For the study of impact ground pressure prevention and control, the first condition is an inherent attribute, and the third condition (creating mining space) is inevitable wherever there are mining activities. Therefore, the focus of the study should be on how to avoid or reduce the degree of load localization of mining surrounding rock, so that it is not enough to cause impact ground pressure. This is what human behavior can do.

The advancement of impact ground pressure monitoring and early warning technology will lay the foundation for improving the efficiency of later impact prevention. At present, the technology of impact ground pressure monitoring and early warning has made significant progress both at home and abroad. The main methods for monitoring and warning of impact ground pressure in our country's coal mines are mine pressure observation method, drill cuttings method, roof dynamic instrument, borehole stress meter, electromagnetic radiation method, geoacoustic method, microseismic method, etc. There are many methods for warning of impact ground pressure, and for different mining areas, one method may be used, or multiple methods may be used for comprehensive monitoring, thus forming different warning modes. The prediction and forecast of impact ground pressure is based on the understanding of the mechanism of impact ground pressure, focusing on the intensity theory and energy conditions of impact ground pressure. In addition to the comprehensive index method of the past, prediction methods can be roughly divided into two categories. The first category of methods mainly includes the drill cuttings method, roof dynamic method, coal and rock body stress measurement method, and rock cake method, mainly used to detect the degree of impact risk in the local area of mining; the second category of methods is system monitoring methods, mainly including microseismic system monitoring method, electromagnetic radiation method, geoacoustic method, and infrared radiation method and other mining geophysical methods, predicting the dangerous state of impact ground pressure based on continuous recording of dynamic phenomena within the rock mass.

Rocks deform and crack under pressure, inevitably releasing elastic energy in the form of pulses, which is transmitted outward in the form of elastic waves, producing an acoustic effect. In mines, ground noise is induced by underground mining activities, with its vibration energy generally ranging from 0 to 103 J; the vibration frequency is high, approximately 150–3000 Hz. Compared to microseismic phenomena, ground noise is a high-frequency, low-energy vibration. Numerous scientific studies have shown that ground noise is a precursor to stress release within the coal-rock mass, and the amount and size of ground noise signals reflect the stress conditions of the rock mass. The ground noise method (acoustic emission) uses the acoustic emission characteristics of the coal body to predict rock burst, and is one of the mainstream prediction methods. Ground noise monitoring is a real-time, continuous coal-rock body acoustic emission monitoring technology that developed rapidly in the late 1970s and early 1980s. The methods used mainly include station-based continuous monitoring and portable mobile ground noise monitoring. The latter is a non-continuous monitoring method, generally used in conjunction with the coal dust drilling method, which can improve the accuracy of rock burst prediction.

During underground coal mining activities, compared to the mining space, the activity of the far-field high-position rock layer is slow, and its rupture radiates a low-frequency ($f < 100$ Hz) high-energy level vibration signal, known as a microseismic event. The microseismic monitoring system detects dangerous sources within a range of about 10 km by responding to the vibration waves emitted by the initiation of coal-rock destruction.

Clearly, the strength of the ground noise signal reflects the energy release process during coal-rock body destruction. The principle of the ground noise monitoring method is to continuously or intermittently monitor the ground noise phenomenon of the rock mass with a microseismometer or seismometer. The tendency of a coal seam or rock mass to impact is judged based on the comparison of the changing pattern of the measured ground noise wave or microseismic wave with the normal

4.1 Key Technologies for Rock Burst Disaster Prevention and Control

wave. The Taozhuang Coal Mine of the Feicheng Mining Bureau in Shandong used a microseismometer to study the rules of rock burst occurrence. The conclusion is: microseismic events increase from small to large, with ups and downs in size, and the frequency and sound are frequent; becoming calm after a group of dense microseismic events is a precursor to rock burst; sparse and scattered microseismic events are normal stress release phenomena, with no risk of impact.

The seismic wave CT method is used to study the continuity of the mined coal seam and reveal the non-uniformity of its hidden structure. The parameter measured is the propagation speed of the seismic wave, which is then used to determine the mine pressure parameters, especially the stress state around the roadway. The advantage of this method is that it is non-destructive, achieves in-situ, one-time, large-scale regional detection underground, and is simple to operate and low in cost.

During the deformation and fracture process under load, the coal-rock body radiates electromagnetic energy outward. The electromagnetic radiation monitoring method is a method to predict coal-rock dynamic disasters using this phenomenon. He Xueqiu, Dou Linming, Wang Enyuan, and others have conducted in-depth research on the theory and technology of electromagnetic radiation monitoring, which has been applied in many mining areas and solved practical problems for many mines.

Other methods for predicting and forecasting rock burst mainly include rock mechanics methods. These methods include the drill cuttings method, coal-rock body deformation measurement method, coal-rock body stress measurement method, geological structure trace analysis method, borehole (bottom) punch squeezing method, rock cake method, etc. Pan Junfeng classified these rock mechanics methods into single manual detection type, comprehensive mine pressure observation type, single physical detection type, and multi-parameter comprehensive monitoring type. The single manual detection type mainly uses one of the methods such as the drill cuttings method, borehole stress monitoring, roof separation observation, and roadway deformation observation. This mode has a large workload for personnel, and the single monitoring result lacks verification and comparison, so the reliability of the warning is the lowest, and it cannot even warn of the occurrence of disasters. The comprehensive mine pressure observation type mainly combines several methods in rock mechanics, such as the drill cuttings method, roof separation observation, roadway deformation observation, borehole stress monitoring, and even combines the monitoring of the support of the mining area and the working resistance of the pillars of the roadway. The single physical detection type mainly uses one of the electromagnetic radiation meter, microseismic monitoring system, ground noise (acoustic emission) monitoring system to monitor and warn of rock burst. The multi-parameter comprehensive monitoring type is a monitoring and warning mode that combines rock mechanics methods with geophysical methods. This mode requires relatively large human and material resources and is the main application mode in typical rock burst mines in China.

Other methods for predicting rock burst include the WET method, elastic deformation method, coal and rock strength and elastic modulus method, drilling powder rate index method, engineering seismic detection method, and so on. The WET method was proposed by the Polish Mining Research Institute for determining the tendency of coal seam bursts. WET is the ratio of elastic energy to permanent deformation energy consumption. The Polish Mining Research Institute stipulates: WET > 5 indicates a strong tendency for bursts; 2 < WET < 5 indicates a weak tendency for bursts; WET < 2 indicates no tendency for bursts. Although this method has some shortcomings, it is basically suitable for China and can be used as one of the indicators for identifying the tendency of coal seam bursts. The elastic deformation method is a method proposed by the former Soviet Union Mining Measurement Research Institute for determining rock bursts. That is, under the condition that the load is not less than 80% of the strength limit, the ratio of the elastic deformation obtained by repeated loading and unloading cycles to the total deformation (K) is used as an indicator of the tendency for bursts. When $K \geq 0.7$, there is a risk of rock bursts. The method of coal and rock strength and elastic modulus is to use the absolute value of the unidirectional compressive strength or elastic modulus of coal and rock as an indicator of the tendency for bursts. This method is relatively simple and is often used as an auxiliary indicator. The boundary value of its indicator must be determined by testing samples from each mine. The drilling powder rate index method, also known as the drilling powder rate index method or drilling inspection method, uses small diameter (42–45 mm) drilling to judge the stress concentration in the rock body based on the amount of drilling cuttings discharged at different drilling depths and its changing rules, and to identify the tendency and location of rock bursts. During the drilling process, if there is a dangerous amount of coal powder measurement or the drill rod is stuck within the specified prevention depth range, it is considered to have a risk of bursts and corresponding measures should be taken. The engineering seismic detection method uses artificial methods to cause earthquakes, detects the propagation speed of these seismic waves, and compiles a relationship chart of wave speed and time. The segment where the wave speed increases indicates a larger stress action, and the tendency of rock bursts is analyzed and judged in combination with geological and mining technical conditions.

It is entirely possible to effectively prevent and control rock bursts by thoroughly studying the mechanism of rock burst occurrence and using scientific prediction methods. The main measures for preventing and controlling rock bursts at present include: adopting reasonable development layout and mining methods, mining protective layers, drilling and blasting pressure relief, hard roof and floor pre-treatment, coal seam pre-water injection, etc.

Reasonable development layout and mining methods are fundamental measures to prevent and control rock bursts. Reasonable development layout and mining methods are extremely important to avoid the formation of high stress concentration and large energy accumulation, and to prevent the occurrence of rock bursts. From the results of comprehensive rock burst classification research, it can be clearly seen that the stress conditions achieved by controlling rock bursts are the key to controlling the occurrence of rock bursts in coal mines. The elastic compression energy that may be released during the advancement of the mining face must be limited to below the range that can cause burst damage. The main principle is: when mining a group of coal seams, the development layout should be conducive to liberation layer mining. When dividing the mining area, a reasonable mining sequence should be ensured to

4.1 Key Technologies for Rock Burst Disaster Prevention and Control

avoid the formation of stress concentration areas such as coal pillars to the greatest extent. The coal mining face of the mining area or panel should advance in one direction to avoid opposite mining and avoid stress superposition. In special parts such as geological structures, a mining procedure that can avoid or slow down stress concentration and superposition should be adopted. The development or preparation tunnels, permanent tunnels, main up (down) hills, main coal chutes and return airways of coal seams with burst danger should be arranged in the floor rock layer or coal seams without burst danger to facilitate maintenance and reduce burst danger. When mining coal seams with burst danger, the longwall mining method of managing the roof without leaving coal to collapse should be adopted. The roof management adopts the full collapse method, and the working face support adopts a retractable support with integrity and protective ability.

When conducting underground mining of multiple coal seams, the mining work of each coal seam affects each other, and the first mining of a coal seam (or layer) can unload the adjacent coal seam for a certain period of time. Therefore, the coordinated mining of the coal seam group should be stipulated at the design stage, first mining the coal seam without burst danger, liberating the coal seam with burst danger, and achieving the effect of reducing the potential danger of rock bursts. Mining protective layers is an effective and fundamental regional prevention measure for preventing and controlling rock bursts.

Drilling pressure relief is a safety measure that uses drilling methods to eliminate or alleviate the danger of rock burst. This method is based on the phenomenon of drilling impact that occurs during drilling construction using the drilling chip method. The closer the drilling is to the high-stress zone, the more energy is accumulated in the coal volume, the higher the frequency of drilling impact, and the greater the intensity, but the amount of coal dust significantly increases during drilling impact. Therefore, a certain fragmentation zone will form around each borehole, and when these fragmentation zones approach each other, they can cause the coal seam to fracture and relieve pressure. The essence of drilling pressure relief is to use the elastic energy accumulated in the coal seam under high-stress conditions to destroy the coal body around the borehole, relieve the pressure of the coal seam, release energy, and eliminate the danger of impact.

Pressure relief blasting is a safety measure that uses blasting methods to reduce the degree of stress concentration in local areas that are at risk of rock burst. Almost all countries in the world use pressure relief blasting as one of the main safety measures when mining coal seams that are at risk of impact.

Thick hard roof easily causes rock burst. One is that the large-area suspension and collapse of the thick hard basic roof of the mining face will cause a high concentration of stress in the coal seam and the roof. The second is that the exposure of rocks near the working face and the upper and lower flat lanes will cause irregular collapses and periodic pressure increases, making it difficult to manage the roof of the working face and maintain the roadway. At present, the more effective treatment methods are roof water injection softening and blasting roof breaking.

The purpose of pre-injecting water into the coal seam is to change the physical and mechanical properties of the impact coal seam through the physical and chemical effects of water, and reduce the impact tendency and stress state of the coal seam. It is currently used in some mining areas.

2. Basic information on the dynamics of rock burst disasters

Controlling rock burst accidents includes controlling the occurrence of rock bursts and possible destructive disasters. The former lies in first understanding the causes and conditions of rock burst occurrence. Based on this, through reasonable "anti-impact" scheme design, including the use of reasonable mining methods, correct mining procedures and mining parameters, etc., the possibility of rock burst occurrence is minimized. The latter lies in the possible time and place of rock burst, through reliable and economical technical measures, including reducing impact energy, reducing the range of impact damage technical measures, as well as safety protection measures for mining faces and roadway workspaces, to prevent casualties and equipment damage and other destructive accidents.

From the results of comprehensive rock burst classification research, it can be clearly seen that the stress conditions for controlling rock burst are the key to controlling the occurrence of coal mine rock burst. It is necessary to limit the elastic compression energy that may be induced and released during the advancement of the mining face to within the range that can cause impact damage. Therefore, when considering the design of the mining plan, attention should be paid to a series of "anti-impact" time–space principles.

(1) Strictly prohibit the arrangement of coal mining lanes and advancing faces in the structural compression stress zone of the original stress field and the peak of mining stress in the mining stress field.
(2) Strive to the maximum extent to achieve excavation and maintenance of the roadway in the stable "internal stress field" (gravity field covered by rock layers that have experienced mining damage).

The relevant information basis for ensuring the design of the mining plan according to the above time and space principles includes:

(1) Information on the distribution of stress in the original stress field that has experienced structural movement damage.
(2) Information on the distribution and development of mining stress under different mining procedures and mining parameter conditions.

The above information must be determined by combining theoretical calculations and actual measurements for specific coal seam conditions and specific mining locations. It is absolutely impossible to use the same empirical data without changing the mining conditions.

In the face where rock burst may occur, the following measures should be taken to control the disaster of the accident.

1. Use the "Underground Rock Layer Dynamic Observation Research Method" (if necessary, supplemented by "Borehole Stress Analysis"), based on obtaining the following related information, to predict and forecast the possible time, place, and intensity of rock burst.

4.1 Key Technologies for Rock Burst Disaster Prevention and Control

(1) The distribution of mining stress, especially the expansion rule of the peak of mining stress in the "internal stress field" range with the advancement of the mining field. Based on this, infer the starting point, end point (with enough "internal stress field" width as a buffer zone) position and the full length of the danger zone. Infer the range of possible impact damage in the advanced roadway.

(2) The laws of basic roof fracturing include the timing, location, and corresponding working face advance distance of the basic lower rock beam, upper rock beam, and hard roof fracturing with the release of bending elasticity. This serves as the basis for predicting the time, location, and possible intensity of roof fracturing induced rock burst. Practice has proven that basic roof fracturing is the main driving force for inducing rock burst in the mining face. The larger the relative stable step of the basic roof, the higher the level of elastic energy accumulated by the roof and coal seam, and the greater the impact strength when the rock burst occurs.

2. In places where high-intensity rock bursts are expected to occur, measures are taken to reduce stored elastic energy and triggering energy, striving to minimize the range of impact damage.
3. In the working face and roadway that can withstand rock burst damage, the correct support method is used to maintain the safety of the working space. For example, the mining face must use stable shrinkable support, and absolutely avoid using unstable support methods such as wooden sheds. Roadways at risk of impact damage should be protected with anchor mesh support as required.

The classification models of rock burst prediction and control research in the excavation face are shown in Figs. 4.9 and 4.10.

Based on existing research, the research team has also proposed a mechanism for evaluating the breeding and unloading energy release of mining stress based on the energy dissipation rate index. The mining and recovery process in coal mines causes the redistribution of surrounding rock stress in the mining space, and the rock mass is induced to cause disasters under the action of dynamic mine pressure, which is one of the main causes of major mine disasters. The occurrence of disasters is closely related to the evolution characteristics of the stress field: the breeding of geostress affects the changes in geological structures, and the rebalancing process of the geostress field affected by engineering disturbances affects the mechanical behavior and physical phase of coal and rock; the effect of unloading and reducing disasters is directly determined by the geostress field. The current conventional dynamic disaster prediction methods are generally divided into non-contact prediction methods represented by microseismic and electromagnetic radiation methods (suitable for large-scale continuous observation of time and space, high cost, strong reliability) and contact prediction methods represented by drilling chip methods (suitable for small-scale non-continuous observation of time and space, low cost, simple operation). However, the unloading characteristics of rock mass mining dynamics in mine dynamic disasters have always lacked scientific and accurate quantitative analysis and expression,

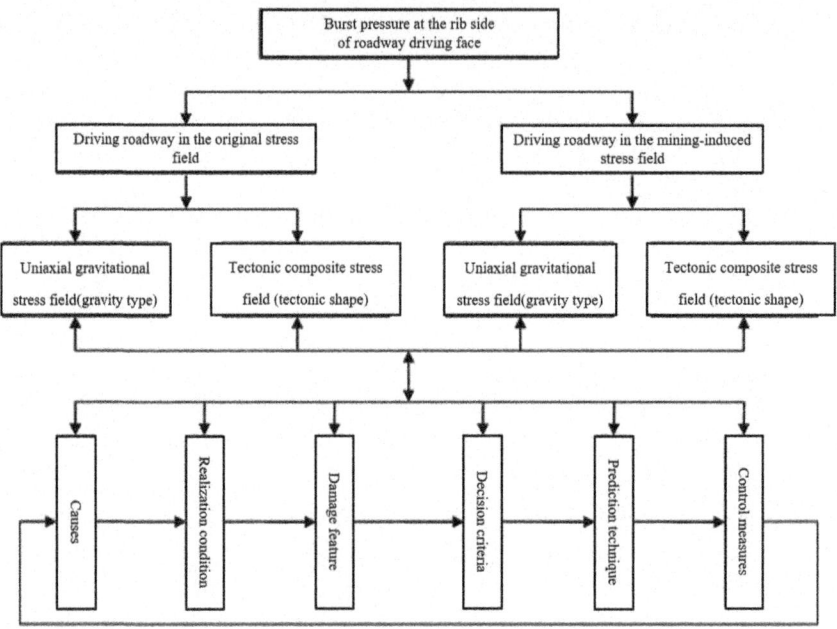

Fig. 4.9 Excavation rock burst classification model

gradually becoming one of the difficult problems that plague the rock mechanics community.

In view of the fact that the occurrence and severity of rock bursts are closely related to the energy stored in the rock mass, which is the result of the combined action of geostress and mining stress. By analyzing the characteristics of mining stress breeding and energy dissipation in the rock mass in front of the working face.

From a theoretical perspective, the definition and expression of the energy release rate are derived, and a preliminary mining stress unloading energy evaluation system that conforms to the actual situation of the site is proposed. The feasibility of monitoring the rock burst based on the energy change of the mining stress energy field is clarified, and a set of dynamic disaster prevention and control quantitative evaluation and control system that can reflect the breeding process of rock burst from the perspective of energy is established in combination with mining stress and the distance from the coal wall. This is of great significance for effectively characterizing the dynamic breeding evolution characteristics of mining stress and the spatiotemporal coupling law, and scientifically quantitatively evaluating the unloading energy release mechanism.

The study of the distribution law of mining stress around the coal mining face, especially in front of the working face, has always been the core content of mining engineering research. The original state of the underground rock mass is not an absolute equilibrium state, but a relative equilibrium. The coal and rock mass is usually regarded as a continuous elastic body for research. The unmined coal and

4.1 Key Technologies for Rock Burst Disaster Prevention and Control

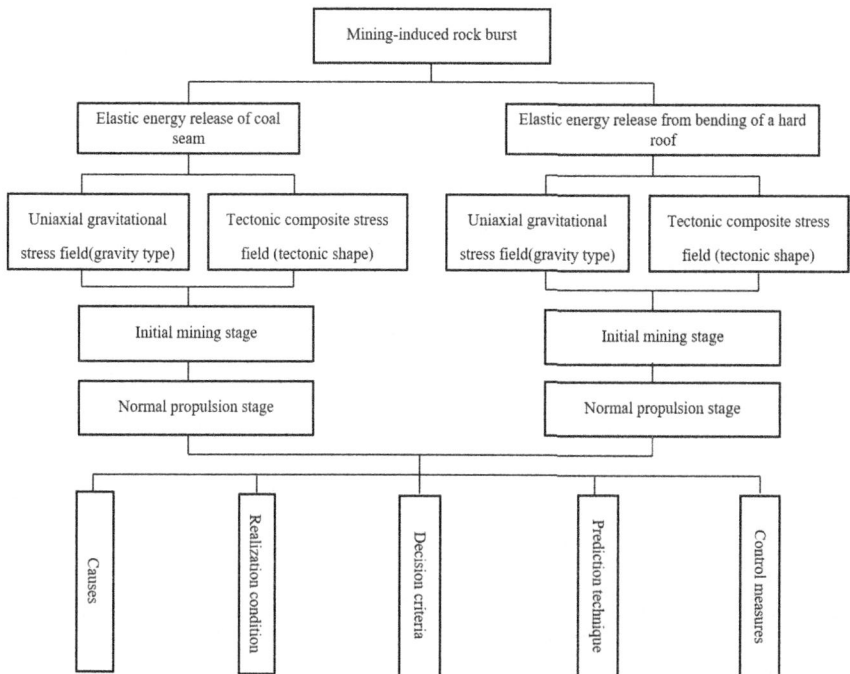

Fig. 4.10 Mining rock burst classification model

rock mass is usually in an elastic deformation state before the excavation of the roadway, and its original vertical stress is equal to the weight of the overlying rock layer. After the underground engineering is excavated, under the action of geostress and the unloading effect of the open space, the balance inside the coal body changes, causing local stress concentration, and the rock mass damage gradually intensifies. If the surrounding rock stress is less than the rock strength, the surrounding rock is still in an elastic state, otherwise, the surrounding rock of the roadway will produce plastic deformation towards the roadway. Energy accumulates in front of the coal wall, and when it reaches the limit conditions of the rock mass, the coal body undergoes rapid dynamic adjustment, that is, it induces a rock burst disaster.

The stress experienced by the rock mass, as the main controlling factor of whether a rock burst occurs, is the starting point for studying the mechanism of rock burst occurrence and the evaluation of stress relief effects based on stress factors. Coal mines infer the state of the coal body in front of the working face and evaluate the stress relief effect by the degree of stress concentration in front of the working face. When the stress in front of the working face is concentrated and the coal wall accumulates a large amount of elastic strain energy, it is very likely to induce a rock burst disaster. After taking stress relief measures, the degree of stress concentration decreases, and the decrease reflects the effectiveness of the stress relief measures. The occurrence of rock bursts is actually the process of "contradiction" between stress and coal rock strength, which is due to the deformation and micro-fracture evolution of the coal rock mass, ultimately leading to macroscopic dynamic instability. It is a process of energy release that is unstable over time and uneven in space. The nurturing of the stress field and the dynamic evolution and development of the energy field during coal mining create conditions for the occurrence and development of rock bursts.

During the advancement of the working face, the transfer and evolution of mining stress are regular. Regardless of the mining method, the rock mass in front of the working face has experienced a complete mechanical process from the original rock stress, axial stress increase (loading) and surrounding rock decrease (unloading) to the failure load. The stress state of the coal rock will change continuously in time and space. The bearing capacity near the coal wall decreases, and the peak of the bearing stress will gradually shift to the inside of the coal wall. The distribution of mining stress on the coal seam will be divided into non-elastic and elastic two intervals. The mining stress in front of the working face changes from the original rock stress zone to the elastic to non-elastic transition. According to the comparison of the stress peak in front of the coal wall and the original rock stress, the stress situation in front of the working face is divided into four parts according to different degrees of stress concentration, namely the stress reduction zone E_{OA}, the stress increase plastic zone E_{AB}, the stress increase elastic zone E_{BC} and the original rock stress zone E_Y, as shown in Fig. 4.11. Among them, the appearance of the stress reduction zone is based on the premise of plastic damage to the coal body, that is, the coal body in the stress reduction zone is in a plastic softening or broken state. The coal body in the stress increase zone is in an elastic state near the basic roof fracture line, still maintaining its own bearing capacity, and the rock mass is relatively intact. The stress increase zone can be further divided into plastic and elastic zones. The boundary between the elastic and plastic zones is the position of the pressure peak, and the stress in the elastic zone monotonically decreases to the original rock stress value.

From a microscopic point of view, take a small cubic unit in the vertical section of the coal body for analysis, as shown in Fig. 4.12. During the retreat mining process of the working face, one side of the boundary coal body unit is empty, the original six-direction force balance state is broken, the initial horizontal stress increases from the empty area side 0 to the original hydrostatic pressure, until the rock mass in the internal original rock stress area recovers to the original state. The vertical stress of the unit body experiences the process of stress reduction to the plastic zone of stress increase to the elastic stress increase zone and then falls back to the original rock stress. The rock mass bears a reduced stress value. As the working face advances, the peak of the bearing stress gradually accumulates and transfers to the interior, creating a new stress peak. When the vertical stress and energy indicators of the unit rock mass in the region reach the limit of the rock mass, the unit produces plastic damage. When the number of damaged unit bodies increases to a certain range, a chain reaction occurs, the pores and cracks between the unit bodies rapidly expand and interact with each other, and the dynamic adjustment speed of the coal body increases to a certain extent, which may induce a disaster.

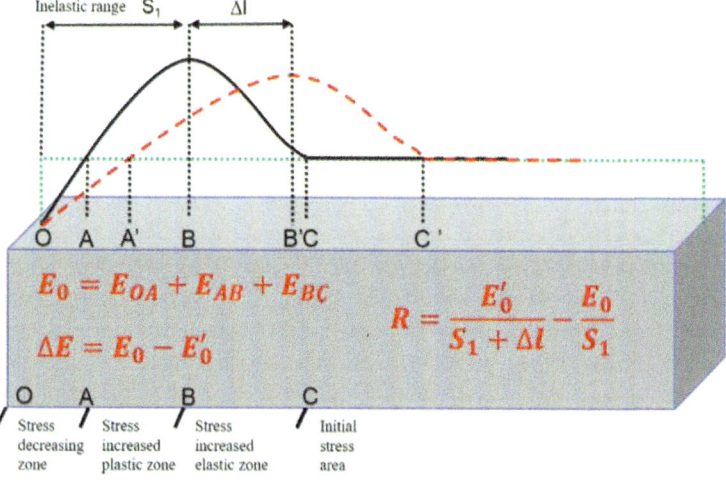

Fig. 4.11 Mining rock mass zoning and energy dissipation model

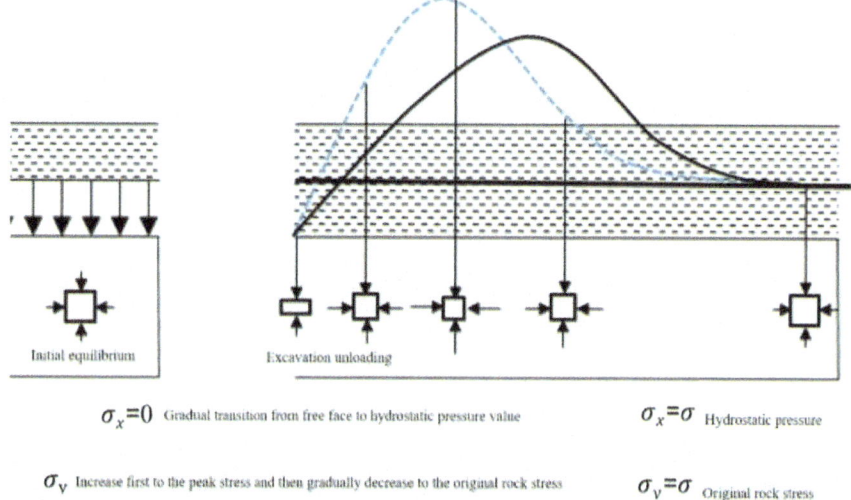

Fig. 4.12 Coal body damage change model

From a macro perspective, because the coal body itself has a certain hardness and strength, when the mining area advances within a certain range, the pressure transmitted by the roof has not reached the limit of coal body damage, the entire coal body is in an elastic compression state, at this time the distribution of mining stress is a monotonically decreasing curve with a peak at the coal wall, at this time the coal wall maintains its elastic bearing capacity, and it is not easy to cause dynamic disaster accidents; as the mining area continues to advance, the pressure transmitted to the coal wall through the roof increases, when the coal wall reaches its elastic bearing limit, it begins to produce plastic deformation, especially under external dynamic disturbances, the change in stress exceeds the bearing capacity of the rock mass and instability occurs. Macroscopically, as the working face advances, the stress state of the coal rock mass in front of the working face is constantly undergoing the conversion process of the stress relief zone, the pressure increase zone, and the original stress zone.

From an energy perspective, the destruction of coal rock mass under mining stress is a process of energy storage and dissipation. Increased energy storage induces rock mass damage, deformation, and destruction, leading to material property degradation and strength loss. When the energy storage reaches the peak limit of rock stress, the energy urgently needs to be released to cause rock mass destruction. Energy release is the essential attribute of rock deformation and destruction, reflecting the continuous development of internal micro-defects in the rock, the continuous weakening of strength, and the final loss. The more severe the damage and fracture, the more energy is dissipated, especially during dynamic destruction, a large amount of elastic energy is dissipated in the form of kinetic energy and so on.

Various types of pressure relief measures are to prevent the elastic strain energy accumulated by the surrounding rock from exceeding the limit storage energy of the rock mass, based on the prevention of shock pressure disaster accidents by reducing the elastic strain energy in the coal body. The construction of an energy dissipation rate index to a certain extent avoids the subjective uncertainty of impact risk assessment purely through stress or energy qualitative, based on the understanding that shock pressure is a destructive phenomenon characterized by energy release and its danger is closely related to the degree of stress concentration and location. Combining stress zoning and energy values to construct an energy dissipation rate index dynamically characterizes the impact tendency in front of the working face and the effectiveness of pressure relief measures. As shown in Fig. 4.11, the formula for constructing a quantitative evaluation of dynamic disaster conditions is defined as:

$$R = \frac{E_0'}{S_1 + \Delta L} - \frac{E_0}{S_1} \qquad (4.7)$$

In the formula, $E_0 = E_{OA} + E_{AB} + E_{BC}$ represents the stress accumulation situation in front of the coal body at a certain moment due to the disturbance of underground engineering, E_0' represents the potential energy of the stress peak coal body accumulated in front of the working face by adopting pressure relief measures, and the energy value of the three zones after the stress in front of the working face is transferred. By comparing the release rate of elastic strain energy in the coal body within the unit range, it can provide a quantitative basis for evaluating the pressure relief effect, $|R|$ the larger the value, the more obvious the overall pressure relief effect. If $R < 0$, it means that the pressure relief effect is achieved, effectively reducing or changing the disaster conditions of the stress field, if $R > 0$, it means that the improper pressure relief disturbance promotes the stability of the stress peak area to decrease, further reasonable and effective measures should be taken to avoid the occurrence of mining disasters. Especially when the mining area is under dynamic load or pressure relief, the scientific and reasonable use of coal rock mass zoning and mining stress energy dissipation situation, incorporating the local stability index of the surrounding rock into consideration, can more accurately reflect the impact danger level of the mining working face, and for each possible impact location, develop corresponding support and pressure relief measures, achieve targeted management of shock pressure, improve the identification and management decision efficiency of shock pressure, which is of great significance for maintaining the safety and stability of the roadway.

4.2 Key Technologies for Pre-control of Roof Water Inrush Disaster

During the excavation or mining process of the mine, the natural balance of the rock layer is destroyed, and the surrounding water body enters the mining working face through faults, water barriers, and weak points of the rock layer under the action of static water pressure and mine pressure, forming mine water hazards. Roof water inrush causes a wide range of roof collapses, crushing major roof accidents of mining supports also occur from time to time. As one of the main disasters in the mining area, coal mine water hazards seriously restrict the safe and efficient mining of coal resources and related water environment control and protection. Water inrush accidents during the advancement of the mining working face can easily cause flooding of its working face and other workplaces, and even cause major accidents of flooding the entire mine [25, 26]. Therefore, in-depth research and development of mine water inrush dynamic disaster warning models and related structural mechanics parameters are of great significance for the safe and efficient mining of coal resources.

4.2.1 Overview of Roof Water Inrush Disaster

1. Causes and Conditions of Roof Water Inrush Disaster

The main cause of roof water inrush accidents is the advancement of the working face causing the fractured rock layer to reach the water-bearing rock layer, especially the water-rich area of the original structural damage.

The conditions that cause accidents mainly include the following points: the existence of water-rich aquifers in the overlying rock layer of the mining area, especially the water-rich areas that have undergone structural damage; the permeability exceeds the drainage capacity of the working face (submerging the working face) or exceeds the drainage capacity of the mine (submerging the mine); a large range of fractured rock layers weaken, slip, and collapse under the long-term soaking of water, crushing

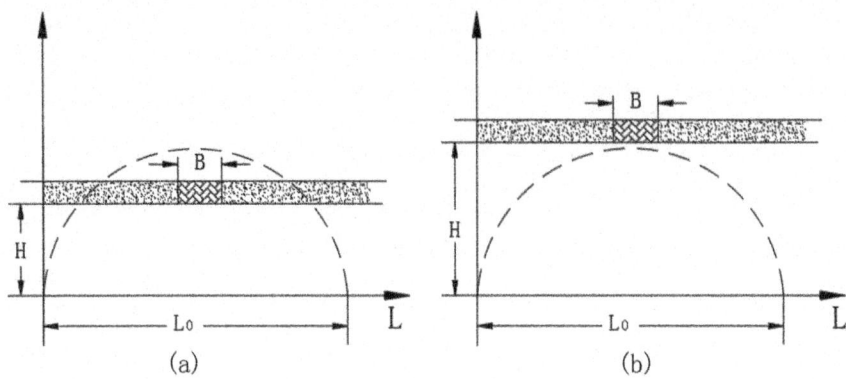

Fig. 4.13 "Fracture arch" and the spatial relationship with the aquifer. H′—Height of the aquifer; B—Width of the aquifer; L0—Length of the working face; L—Length of the working face advancement

the support; the stagnation time of the working face after water permeation exceeds the limit, and the working face is flooded by water permeation. The shear strength of fractured rock layers and fracture biting surfaces decreases due to prolonged soaking.

The fractured rock layer that subsides with the advancement of the mining area communicates with the water-rich aquifer. This includes both entering the aquifer and direct contact, as shown in Fig. 4.13.

2. The relationship between mining stress and roof water disaster

Coal mine outburst and water inrush accidents refer to events in which water of different forms and sources enters the mine through certain channels during the construction and production of coal mines, bringing adverse effects and disasters to coal mine construction and production. The formation and occurrence of coal mine water inrush accidents are based on specific environments and conditions. The occurrence of either floor water outburst or roof water inrush accidents cannot be separated from the three necessary conditions of water inrush source, water inrush channel, and water inrush intensity. The existence or absence of the source of water inrush, the channel of water inrush, and the intensity of water inrush determines whether a coal mine water inrush accident occurs, and different combinations of the three will produce different types of coal mine water hazards. The water inrush source is the root cause of mine water inrush accidents. If there is no water inrush source, water inrush accidents will not occur. After coal seam mining, under the action of mining stress, the top and bottom plates of the working face develop cracks, the overlying rock is damaged, and the "fracture arch" continues to develop upward. At this time, if the mining stress and the "fracture arch" affect the floor pressurized water source or the roof aquifer, especially the original structurally damaged water-rich area, and the water inrush intensity exceeds the mine's drainage capacity, a roof water inrush accident will occur, causing casualties and property losses, and in severe cases, it may even flood the entire mine.

4.2 Key Technologies for Pre-control of Roof Water Inrush Disaster

The original stress field, mining stress field, and seepage field of the surrounding rock in the mining area together constitute the controlling factors of the top and bottom plate water inrush. At the same time, with the increase in the mining depth of the working face, the impact of the mining stress field becomes more significant year by year. When the coal seam is not mined, the voids, cracks, etc. are in a closed state, and the water in the aquifer is in a relatively balanced state. After the coal seam is mined, the roof is exposed, the surrounding rock changes from three-way compression to two-way or one-way compression, and even some rock bodies appear tensile stress. During the mining process, every point of the floor rock layer has undergone a "compression → stress relief → recompression" process. It is these stresses that cause the change in the fracture rate of the floor rock layer. The development of mining stress makes the coal rock body fracture conductive development, greatly increasing its permeability, the water outburst channel is conductive, leading to water inrush events.

For nearly horizontal coal seams, when the retreat working face advances and the fracture zone affects the aquifer, the water inrush volume Q can be approximately calculated by formula 4.8 as shown in Fig. 4.14.

$$Q = 2(Q_1 + Q_2) \tag{4.8}$$

where, Q_1 is the water gushing from the fracture surface in the advancing direction when the aquifer first fractures; Q_2 is the water gushing from the fracture surface in the length direction of the working face when the aquifer first fractures.

The water gushing from the fracture surface in the advancing direction can be calculated by the following formula:

$$Q_1 = k \cdot \omega_c \cdot I \tag{4.9}$$

where, k is the permeability coefficient of the aquifer, that is, the flow rate of water through the unit water-crossing section when the water pressure slope $= 1$. The reference values of the permeability coefficients of rock layers with different particle sizes and porosities obtained from actual measurements are shown in Table 4.1.

ω_c is the water outflow cross-sectional area of the aquifer in the advancing direction. Under the known aquifer thickness m_0 condition, this cross-sectional area is determined by the first fracture step distance C_{oH}, that is:

$$\omega_c = C_{oH} \cdot m_0$$

The water pressure gradient I value of the closed aquifer is expressed by the following formula, that is:

$$I = \frac{dm_0}{dx}$$

Substitute into Eq. 4.9 to get:

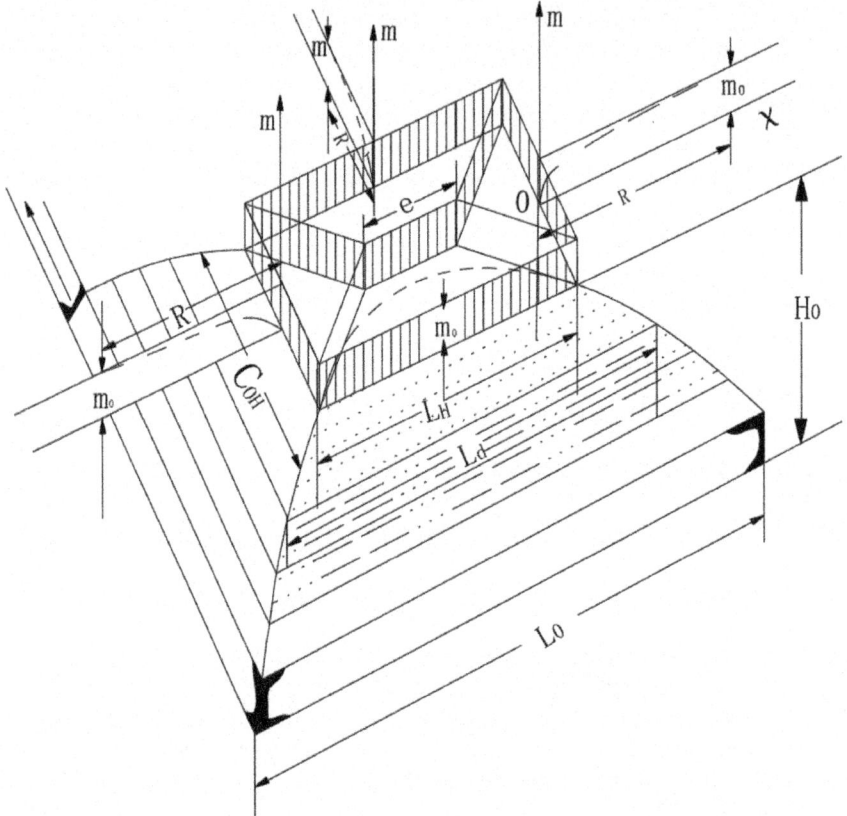

Fig. 4.14 Comparison chart of predicted water inrush volume

$$Q_1 = k \cdot C_{oH} \cdot m_0 \cdot \frac{dm_0}{dx} = C_o \cdot q$$

where, q is the water inflow per unit length of the overwater cross-section, its expression is:

$$q = k \cdot m_0 \cdot \frac{dm_0}{dx}$$

Therefore, we have:

$$\frac{q}{k} dx = m_0 \cdot dm_0$$

Relative to the water inflow impact range of the aquifer and integration between the two fractures:

4.2 Key Technologies for Pre-control of Roof Water Inrush Disaster

Table 4.1 Reference values of permeability coefficients of water-bearing layers

Permeability layer classification	Permeability coefficient	Related rock layer types
Highly permeable rock layer	$k > 10 m/d$	Coarse sandstone, conglomerate, karst developed rock layers
Permeable rock layer	$k = (10 - 1)m/d$	Sand, fracture developed rock layers
Slightly permeable rock layer	$k = (1 - 0.01)m/d$	Silty sand, weakly fractured rock layers
Permeability layer classification	Permeability coefficient	Related rock layer types
Extremely weakly permeable rock layer	$k = (0.01 - 0.001)m/d$	Silty clay, clay, silt
Impermeable rock layer	$k < 0.001 m/d$	Dense hard rock layers (including hard sedimentary rocks, igneous rocks, metamorphic rocks), clay, silt, mudstone

$$\int_0^r \frac{q}{k} dx = \int_0^{m_0} m_0 \cdot dm_0$$

Therefore, we get:

$$q = k \cdot \frac{m_0^2}{2R}$$

From this, we can find Q_1 is:

$$Q_1 = k \cdot C_{oH} \cdot \frac{m_0^2}{2R} \tag{4.10}$$

where, k is the permeability coefficient; m_o is the thickness of the aquifer; C_{oH} is the fracture step distance of the aquifer; R is the water inflow impact radius of the aquifer; R is the water inflow impact distance of the aquifer fracture (i.e., the distance from where its water level begins to drop to the fracture outflow surface). If this range is unknown, drilling into the aquifer can be used for measurement.

As can be seen from Fig. 4.15: If the pressure measuring hole is at a distance R_i from the water outflow cross-section, the measured water pressure is m_i, and $m_i < m_0$, then it is not difficult to find the impact distance R value is:

$$R = \frac{m_0}{m_i} R_i \tag{4.11}$$

Fig. 4.15 Aquifer fracture distance value

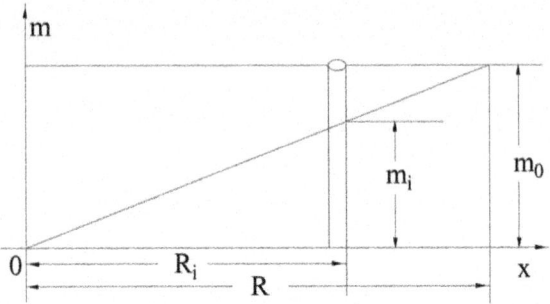

Similarly, the water inflow amount Q_2 in the working face length direction can be derived as:

$$Q_2 = k \cdot L_H \cdot \frac{m_0^2}{2R} \tag{4.12}$$

In Eq. 4.12 L_H is the first fracture line length of the aquifer in the working face length direction. It depends on the length of the mining working face (L_0) and the height of the aquifer from the mined coal seam (H_0). Given that in most cases, the height of the fractured rock layer is approximately half the length of the working face. Therefore, the fracture boundary of the fractured rock layer advancing in the mining area can be approximately expressed as a semi-circular arch as shown in Fig. 4.16. From this, the expression for L_H, the length of the first fracture line of the water-bearing layer when the known height of the water-bearing layer is H_0, can be approximately obtained:

$$L_H = L_0 - 2X_i \tag{4.13}$$

where

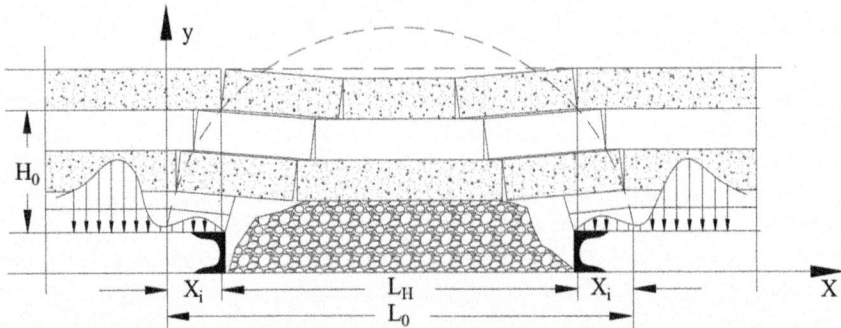

Fig. 4.16 Relationship between the length of the water-bearing layer fracture line and the length of the working face

4.2 Key Technologies for Pre-control of Roof Water Inrush Disaster

$$X_i = \frac{L_0}{2} - \sqrt{\left(\frac{L_0}{2}\right)^2 - H_0^2}$$

Therefore, we get:

$$L_H = \sqrt{\left(\frac{L_0}{2}\right)^2 - H_0^2}$$

The above formula shows that the permeability of the water-bearing layer increases with the increase of the length of the working face and decreases with the increase of the height of the water-bearing layer from the mined coal seam.

The permeability of the water-bearing layer when the fracture zone of the inclined coal seam affects the water-bearing layer is expressed by formula 4.13:

$$Q = 2(Q_1 + Q_2) \tag{4.14}$$

where, Q_1 is the water inflow on the first fracture surface of the water-bearing layer in the direction of the working face advancement. It can be approximately estimated by the following formula:

$$Q_1 = k \cdot C_{oH} \cdot m_0 \cdot I \tag{4.15}$$

where, I is the hydraulic slope, and other symbols are the same as before. The hydraulic slope of the inclined coal seam is determined by the inclination angle of the coal seam, that is:

$$I = tg\alpha$$

Therefore, the expression for Q_1 is obtained:

$$Q_1 = k \cdot C_{oH} \cdot m_0 \cdot tg\alpha \tag{4.16}$$

Q_2 is the water inflow on the fracture surface of the water-bearing layer in the direction of the working face length, and its expression is:

$$Q_2 = k \cdot L_H \cdot m_0 \cdot tg\alpha \tag{4.17}$$

where

$$L_H = \sqrt{\left(\frac{L_0}{2}\right)^2 - H_0^2}$$

The conditions for the occurrence of roof seepage accidents can be expressed by the following formula:

$$(Q - Q_B)T \geq V_{max} \tag{4.18}$$

where, Q is the roof seepage flow rate (m³/h); Q_B is the capacity of the drainage equipment (m³/h); T is the time of work stagnation (hours); V_{max} is the maximum water storage capacity of the possible accident site (m³).

From formula 4.18, it can be seen that on the basis of correctly predicting the possible location of roof seepage accidents and the possible seepage flow rate (Q), preparing sufficient drainage equipment capacity (Q_B) and ensuring the rapid advancement of the working face are key to preventing water seepage accidents at the mining face. For mines, correctly predicting the number of advancing faces based on the mine's production capacity, predicting the maximum water inflow, ensuring the capacity of the drainage equipment and sufficient water storage capacity are key to preventing mine water seepage accidents.

4.2.2 Mechanism of Roof Seepage Disaster

From the late 1970s to the 1980s, China's research on coal mining under water bodies began to explore the formation mechanism of water-conducting fracture zones from the perspective of rock mass structure control theory, focusing on the deformation and failure process of overburden, influencing factors, and prevention and prediction of roof and floor water. In the late 1980s and into the 1990s, it was recognized that the water barrier layer is composed of different rock layers with small thickness, and only the key water barrier layer in it can well meet the thickness-length ratio conditions of beams and slabs. The key layer theory, which plays a key role in the water barrier layer, was proposed. Academician Qian Minggao's key layer theory believes that the separation layer is generated between layers as the overburden strata sink, and the development, evolution and spatiotemporal distribution of the separation layer are controlled by the key layer and develop under the key layer. As the working face advances, the space volume of the separation layer continuously increases before the key layer first breaks, and after the first break, the separation layer located in the middle of the goaf is gradually compacted, while the separation layer area on both sides of the goaf is continuously moving forward with the advancement of the working face, and its height and width are only 1/4–1/3 of that before the key layer first breaks. In 1981, Liu [27] The concept of overburden damage was first proposed in China, suggesting that after longwall mining, the deformation characteristics and water conductivity of the overburden will divide the overlying rock layer into "three zones". Currently, this theory is mainly used as the basis for studying the water seepage mechanism of the roof. Later, Professor Gao Yanfa broke through the traditional concept of "three zones" and proposed a "four zones" model of rock movement [28], dividing the bedrock according to its mechanical structure characteristics after damage into fracture zone, detachment zone, bending zone, and loose alluvial layer zone, further broadening the understanding of the roof seepage mechanism. It is believed that after coal seam mining, although the rock layers above the collapse zone are fractured or subsided, they still maintain a layered structure, but they no longer have continuity along the direction of the stratum plane. When the fracture damage phenomenon develops to a certain extent, the overlying strata will show detachment and subsidence. Academician Song Zhenqi of the Chinese Academy of Sciences and others [25] proposed the theory and related information basis for predicting and controlling coal mine water disaster accidents, providing a basis for the informatization of safe and efficient mining and management of coal mines. For type I and type II detachment water disasters, the principle of explosion is used to describe the sudden fracture of the hard rock above the detachment, which instantly generates impact pressure on the detachment water body. After the detachment water body obtains the initial impact pressure, it propagates downward to the rock body below the detachment in the form of shock waves. This causes the lower rock layer to generate stress waves under the impact, thereby generating cracks and seeping water after expanding and communicating with the water-conducting cracks; for type II detachment water disasters, according to the theory of solid support beams, the formula for determining the limit thickness of the rock body below the detachment is used, and the thickness calculation formula for the soft rock layer to break during the first pressure and periodic pressure is derived.

Although great progress has been made in the research on the mechanism and prevention of roof seepage, there are still many problems that need to be further expanded and developed. Such as the mechanism of roof seepage under different geological and mining conditions, the evolution of fractures in water-rich overburden roofs, the law of destruction and the law of potential water seepage in overburden, the development height and change law of water-conducting fractures, etc. In recent years, with the implementation of comprehensive top coal caving mining technology, it is necessary to further study the movement of overburden and the mechanism of roof floor water and gas outburst under comprehensive caving mining conditions.

4.2.3 Dynamic Information Basis for the Prevention and Control of Roof Seepage Disasters

The hydrogeological conditions of mines are becoming more complex, the factors affecting water inrush are increasing, and the mechanisms and types of water inrush are complex and varied. Article 3 of the "Coal Mine Water Prevention and Control Regulations" proposes a new "sixteen-character" basic principle for mine water prevention and control in our country, namely "prediction and forecasting, doubt must be explored, exploration before excavation, treatment before mining". Based on a deep analysis of the mechanism of floor water inrush, our country's water prevention and control workers have explored a series of ways and methods suitable for the management of different types of water inrush disasters, including drainage and pressure reduction, curtain grouting, grouting reinforcement, local drainage, pressure mining, and mine water prevention and drainage. Among them, "drainage and pressure reduction" and "grouting to transform the floor and aquifer" are two important engineering measures for the prevention of coal mine water disasters. Drainage and pressure reduction often control the source of water inrush through deep drainage and strong drainage. Although it can eradicate water disasters from the source and ensure safety, it destroys a large amount of water resources and karst deposits, and the technology is expensive and difficult to achieve when the water supply is abundant and the replenishment source is sufficient, so it is less used. "Grouting to transform the floor and aquifer" is a method of water prevention and control that developed in the mid-to-late 1980s. The basic principle is that under a certain time and pressure, the grout is injected into the grouted layer, and the grout dehydrates and solidifies in the original water-occupied pores or channels to form a water-blocking cemented body, thereby improving the water-separation performance of the floor rock layer, reinforcing the floor water-separation layer, and enhancing the floor's resistance to destruction.

1. Key technologies for the prevention and control of roof seepage disasters

Advanced water drainage in the mining face is a traditional underground water prevention and control technology, but when the hydrogeological conditions and water filling factors of the mine cannot be ascertained from the ground work, it still has very important use value in predicting and avoiding the sudden occurrence of major water disaster accidents. The means of advanced water drainage in the general mining face include geophysical exploration, chemical exploration, drilling, and pit exploration, among which geophysical exploration and drilling are the main commonly used means. The concept of advanced water drainage in the mining face is that direct means such as drilling or pit exploration must be used for water drainage, and indirect means such as geophysical or chemical exploration are not allowed to directly explore water.

Underground geophysical prospecting technology has made significant progress, basically covering all aspects of underground mining engineering, and has initially formed a relatively complete system of underground geophysical prospecting technology. In terms of advanced detection at the excavation face, the main methods include direct current method, transient electromagnetism, etc., but the commonly used methods with better application effects are mainly direct current method and transient electromagnetic method. The former has the advantage of no detection blind spots and is a contact detection, but the disadvantage is that it can only detect in one direction towards the excavation in the underground tunnel space, with a short detection distance (60–80 m), and a relatively large volume effect; the latter has the advantage of multi-directional detection in the underground tunnel, with a relatively long detection distance (80–120 m), and a relatively small volume effect, but its disadvantage is that there is a blind spot of about 20 m in the detection results, and it is a non-contact detection. In terms of advanced detection at the mining face, the main

4.2 Key Technologies for Pre-control of Roof Water Inrush Disaster

detection methods include frequency electrical perspective, transient electromagnetism, radio perspective CT, slot wave seismic and elastic wave CT tomography, etc., among which frequency electrical perspective and transient electromagnetic method mainly detect the water-bearing layer of the coal seam roof and floor, the water-richness of burnt rocks and the distribution of old empty water areas and water-conducting structures, etc., radio perspective CT, slot wave seismic and elastic wave CT tomography mainly detect coal seam structure, coal thickness changes, magmatic rocks and fire areas, etc. The new geophysical prospecting technologies currently used underground also include borehole peeping, hole tunnel perspective, borehole inclinometer, hole depth detection and hole opening orientation, etc. Underground directional drilling, branch hole making technology and advanced directional drilling water drainage technology at the mining face. Drilling ahead for water drainage is a traditional method that is still widely used in China and is one of the most important and effective methods for underground water control. The large amount of advanced drilling work seriously affects the excavation speed of the excavation face, and there is a sharp contradiction between advanced water drainage and excavation speed, which seriously affects the promotion and application of advanced drilling water drainage technology underground. Therefore, it is particularly important to reduce the number of water drainage drill holes while ensuring the quality and effect of advanced water drainage. Underground directional water drainage drilling has many advantages such as small hole opening density, high controllability of drilling trajectory, multi-directional in one hole, less ineffective footage, can work in parallel with drilling without affecting each other, accurate detection target point and high drainage efficiency, and has a good application prospect in the detection and treatment of adverse geological anomalies such as geological structure and coal thickness changes, in the directional grouting reinforcement of the water-bearing layer at the bottom of the mining face and the transformation of the water-filled water-bearing layer, in the advanced exploration and drainage of groundwater and old empty water in the water-filled water-bearing layer of the roof and floor, etc. Activation of mine geological structure under mining conditions, height of fracture of coal seam roof and floor, and water hazard monitoring and early warning technology. Under mining conditions, the formation and occurrence of mine water hazards have a process of gestation, development and occurrence. At different stages of this process, corresponding signs of water inrush and water penetration will be released in terms of stress and strain (especially at geological structure locations), water pressure (water level), water inflow, water chemistry and water temperature, etc. Timely, accurate and effective monitoring of these signs, and establishing a complete system that integrates mine water hazard monitoring, identification and early warning technology, has important theoretical significance and practical value for preventing major and particularly serious water hazard accidents. At present, the high-precision microseismic monitoring technology that can diagnose and display in real time, dynamically, and in a planar manner the deformation and destruction process of the coal seam roof and floor, the activation intensity and intensity of faults and subsidence columns and other geological structures, and related spatiotemporal parameters under mining conditions, has good application effects in China.

Domestic and foreign experts and scholars have conducted extensive research on the technical measures for the control of roof seepage disasters. Techniques such as off-layer grouting filling are adopted to solidify and control the surrounding rock mass of the working face. Off-layer is one of the important reasons for the formation of the main cracks in the roof. The technology of overburden off-layer grouting is based on the understanding of the spatiotemporal rules of the formation and development of off-layer in the overburden layer above the coal seam. By drilling from the ground and injecting grouting material into the injectable off-layer band of the overburden body under high pressure, the off-layer space is filled, the degree of overburden damage and the concentrated development of mining fractures are reduced, the formation of dominant water-conducting fractures is eliminated or alleviated, and the subsidence of the ground surface is slowed down. With the increase of mining depth and the mining of the lower group of coal, mine water disasters occur frequently, potential water hazards intensify, and the overall or local grouting technology on the ground and underground shows obvious advantages in quickly blocking and controlling water inrush disasters and eliminating potential water hazards, such as curtain cut-off grouting in the concentrated recharge zone and strong runoff zone of underground water in mines, local pre-grouting transformation of water-filled aquifers and blocking of water-conducting channels, directional grouting on the ground to build a "water-blocking plug" on the upper part of the karst collapse column to cut off the deep supply water source channel, and the comprehensive grouting technology and technology on the ground to establish a "water-blocking wall" in the water-filled roadway, etc. These sets of coordinated grouting technologies, rapid directional drilling and branch drilling technologies, different technologies and methods of ground and underground grouting, various grouting water-blocking effect evaluation methods and criteria, etc., provide a strong technical guarantee for the prevention and control of mine water disasters.

Comprehensive mechanized filling coal mining technology. This technology uses coal walls, supports, and fillings for uninterrupted relay support of the direct roof, restricts the deformation of the direct roof, transforms the direct roof into the basic roof, changes the rock layer of the mine pressure beam transmission, controls the mine pressure manifestation, stabilizes the water-bearing layer structure of the coal seam roof and floor, thereby achieving the purpose of preventing and controlling water disasters of the coal seam roof and floor. In addition, comprehensive mechanized filling coal mining technology also has important significance in controlling ground subsidence, eliminating mine ground gangue hills, preventing gas accumulation in mined-out areas, eliminating the hidden danger of spontaneous combustion of coal seams, and reducing geological problems in the mine environment.

Reasonably adjust the advancement speed. Mining disturbance causes the stress redistribution of the surrounding rock of the working face. Under the condition of comprehensive mining, the advancement speed of the working face also has an important impact on the mechanical environment of the surrounding rock. Within 20 m in front of the comprehensive working face, when the advancement speed of the working face is not large, although the stress of the surrounding rock is not high, the deformation rate of the roadway increases sharply; with the increase of the advancement speed of the working face, the area of the stress reduction zone and the rock mass destruction zone around the working face decreases, the peak stress in front of the working face approaches the working face, and the displacement of the rock mass around the working face decreases. It is proposed that appropriately increasing the advancement speed of comprehensive mining is beneficial to the management of the working face and safe production. In actual mining, the relationship between the advancement speed and safe production and the maximization of production benefits should be considered comprehensively and determined in combination with various factors, and a reasonable advancement speed should be proposed to ensure pressure balance and not allow intermittent mining.

Determine the reasonable width and mining height of the working face. According to the simulation results of the deformation, destruction and migration rules of the upper roof during the comprehensive mining process with the first mining working face length of 150 m, the mining height of the working face is relatively large, the mine pressure manifestation of the overlying rock layer on the working face is severe, and it has obvious pressure characteristics. The development height of the water-conducting fracture zone is about 14–15 times the mining height; when it advances to about 290–300 m, the overlying rock layer is widely affected by mining and produces caving and migration, and the fracture height reaches the sandstone water-bearing layer of the Luohe Group, providing a possible channel for seepage from the water-bearing layer. Therefore, in the normal mining process of the working face, different technical measures should be taken for different stages.

Optimize roadway layout and adjust mining technology. Studies have shown that the mining height of the comprehensive mining working face is relatively large, the mine pressure manifestation of the overlying rock layer on the working face is severe, and it has obvious pressure characteristics. During the advancement of the working face, the periodic fracture of the overburden and the effect of the advanced mining stress in front of the working face always form a through-fracture development zone near the working face, and the fractures connect with each other to form several possible water-conducting channels; the caving rock blocks in the middle of the mined-out area behind the working face gradually compact, causing the fractures to squeeze and close, forming a local fracture closure zone. At the same time, it is easy to form a speckle line along the down-flat roadway of the working face, and the speckle line is the main channel for the water in the water-bearing layer to enter the working face.

Improve the drainage system and hydrological observation system. Improve and transform the mine water prevention and drainage system, and timely formulate safety plans such as increasing the drainage capacity of the existing equipment in the mine, to improve the comprehensive disaster resistance ability.

Workplace safety management. The hydrogeological conditions of the working face are extremely complex, requiring the construction unit to strictly implement and carry out the mining operation procedures, special water prevention measures, and water disaster emergency treatment and prevention plans for this face. Before mining, drills for water disaster emergency treatment and prevention plans should be conducted. Workers at the working face should be proficient in mastering the signs of water inrush and the route to avoid water disasters, and should report and take measures in a timely manner according to the water disaster emergency treatment and prevention plan and operation procedures, special water prevention measures, and evacuation, etc.

High-power, high-lift, large-flow mine submersible electric pump technology. In the past, the mine submersible electric pumps that were always dependent on imports were only used for emergency rescue drainage after accidents due to their high prices. However, with the progress of science and technology in our country, the high-power, high-lift, large-flow submersible electric pump technology with independent intellectual property rights has gradually matured. A large number of high-quality and inexpensive domestic mine submersible electric pumps have provided material guarantees for the widespread use of water disaster prevention and control in our country's mines. They have provided technical support for the difficult problem of setting up waterproof gates around the mine bottom car yard under complex and extremely complex hydrogeological conditions and the conflict with the original rigid technical standards required by the state. They have also provided an alternative technical choice for mines with complex and extremely complex hydrogeological conditions that really do not have the conditions to build waterproof gates.

The prediction and control of water seepage accidents in coal mine roofs are crucial. From the perspective of inducing factors and implementation conditions, coal mine roof water seepage accidents are traceable, requiring scientific and accurate prediction and the adoption of advanced control technology. Only in this way can the probability of roof water seepage accidents be reduced, the harm caused by accidents be reduced, and production safety be ensured.

2. Basic information on roof water seepage disasters

During the advancement of the working face, roof seepage and floor water inrush can both potentially cause major accidents such as flooding of the working face and workplaces along the water flow path, or even the entire well. Large-scale roof collapses caused by roof seepage, which crush the supports in the mining area, also occur from time to time. Under the given mining conditions, when the aquifer is parallel to the coal seam, the model and related criteria for judging the possibility of seepage are shown in Fig. 4.17. When the advancement length and the working face are equal in length, the range of damage to the overlying rock layer no longer continues to increase, and the height of the "fracture arch" reaches its maximum. If the "fracture arch" formed under this working face length condition communicates with the upper aquifer, a roof seepage accident is inevitable. On this basis, a seepage possibility prediction model (Fig. 4.18) is established, and seepage criteria (Table 4.2) are proposed based on the analysis of the roof seepage possibility prediction model.

For the first mining face, the relationship between the developed damage arch and the working face length (L) is obtained as follows:

$$h = K_h L ..... K_h = 0.5-0.7$$

4.2 Key Technologies for Pre-control of Roof Water Inrush Disaster

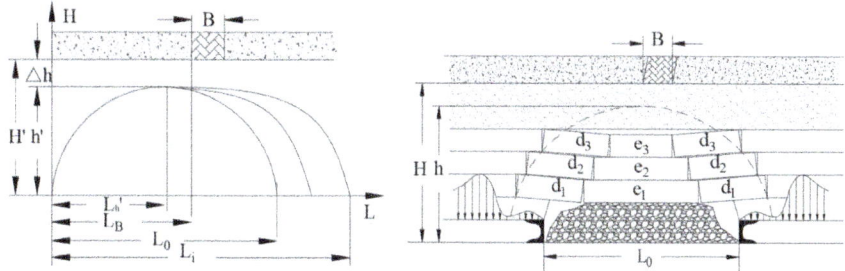

L is the working face advancement step; L_0 is the working face length; h' — the height of the fracture arch; L_h' is the distance from the arch center to the cutting eye; L_B is the distance from the aquifer to the cutting eye; L_i is the working face advancement distance; B is the width of the water-rich area; h is the development height of the fractured rock layer; H is the height of the water-bearing rock layer.

Fig. 4.17 Roof seepage prediction structure diagram

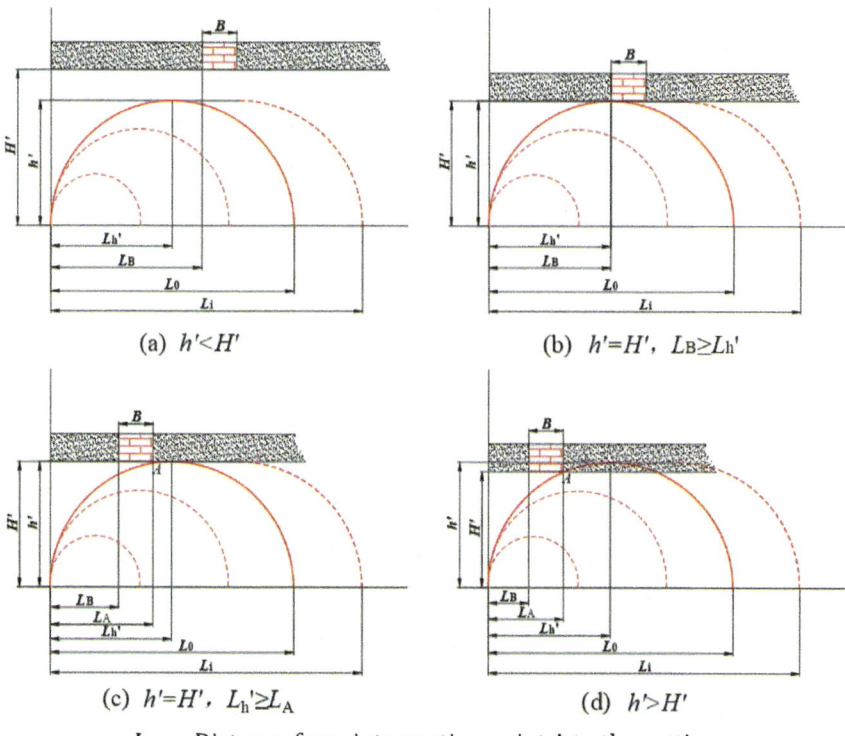

(a) $h' < H'$

(b) $h' = H'$, $L_B \geq L_h'$

(c) $h' = H'$, $L_h' \geq L_A$

(d) $h' > H'$

L_A — Distance from intersection point A to the cutting eye

Fig. 4.18 Roof seepage possibility prediction structure diagram. L_A—Distance from intersection point A to the cutting eye

Table 4.2 Roof seepage criteria

Model number	Judgment criteria	Judgment result
(a)	$L_B > L'_h$	Impermeable
	$L_B < L'_h$	Impermeable
(b)	$L_B \geq L'_h$	Permeable
	$L_B < (L_A - B)$	Impermeable
(c)	$L_B > (L_A - B)$	Permeable
	$L_B < (L_A - B)$	Impermeable

The step distance (L_0) and center distance (L_h) when the working face advances to the damage arch height are as follows:

$$L_0 = L \quad L_h = 0.5L$$

From the structural diagram predicting the possibility of roof water penetration in Fig. 4.18, it is known that when the distribution curve of the bottom of the water-bearing layer and the edge curve of the fracture arch in the direction of the working face advance are known. By solving simultaneously, the position of the intersection point and the distance from the working face to the cutting eye can be obtained L_A. If the water-bearing layer is basically parallel to the coal seam in the direction of advancement, and the strength difference of the intermediate rock layer is not large and the deposit is stable. Then the approximate equations of the related curves are:

Bottom curve equation of water-bearing layer:

$$y = ax + B = H \tag{4.19}$$

Edge curve equation of fracture arch:

$$y = \sqrt{h^2 - (h-x)^2} \tag{4.20}$$

When the fracture arch wave reaches the water-bearing layer (i.e., $h \geq H$), the intersection point (A) is at a distance from the mining eye (L_A), which can be obtained by solving Eqs. 4.19 and 4.20 simultaneously:

$$L_A = \frac{2h \pm \sqrt{4h^2 - 4H^2}}{2} \tag{4.21}$$

where: $h = 0.5L$, thus obtained:

$$L_A = \frac{L \pm \sqrt{L^2 - 4H^2}}{2} \tag{4.22}$$

4.2 Key Technologies for Pre-control of Roof Water Inrush Disaster

Obviously, when the width of the water-rich area of the water-bearing layer is, the minimum distance from the cutting eye to the edge of the water-rich area $L_{B\min}$ is:

$$LA_{B\min} = L_A - B \tag{4.23}$$

Similarly, when the position of the water-rich area of the water-bearing layer is known from the cutting eye, the maximum allowable working face length that does not cause water penetration accidents can be inversely calculated $L_{\max}$ is:

$$L_{\max} = \frac{H^2 + (L_B + B)^2}{L_B + B} L_{\max} \tag{4.24}$$

The key to controlling the possibility of water penetration is to understand the position of the water-bearing rock layer and the range of its water-rich area (B). Based on this, by adjusting the relationship between the length of the working face and the position of the cutting eye relative to the water-rich area, it is ensured that the fracture rock layer does not reach the water-bearing rock layer, especially the water-rich area during the entire process of the working face advancement.

The information for predicting and controlling roof water penetration accidents includes four aspects: water source information, structural movement damage situation, mining roof movement damage information, and mining stress distribution information. (1) Water source information, including: the number of water-bearing layers in the roof, position, thickness and water-bearing characteristics (including water properties, area, distribution of water-rich areas, water pressure and replenishment water source conditions, etc.); position and thickness of the roof water barrier layer; roof folding, fault. (2) Structural damage situation, including: folding damage situation and fault damage situation, it is the structural damage that connects the water-bearing layers and replenishment water sources, forming a water-rich area. (3) Mining roof movement damage information, including the movement damage situation of the overlying rock layer under different working face lengths and the development change law with the advancement of the mining area. The key information includes: the range of the overlying rock layer entering the damage under a given working face length, including the thickness of the direct roof, the basic roof thickness, the height of the water-conducting fracture zone, etc.; the movement step of each rock beam in the direct roof water-conducting fracture zone under the action of gravity, especially the first fracture step, etc.; the water-bearing layer that may enter the water-conducting fracture zone, the first fracture and periodic fracture step under the action of gravity. (4) The focus of the mining stress distribution information is the width of the "internal stress field" of the coal wall entering the damage around the mining area. It is a reliable interval for arranging the prediction of the basic roof pressure measurement means, and it is also the basis for predicting the time and place where the roof water penetration may occur.

4.3 Key Technologies for Roof Disaster Prevention and Control

Compared to gas and water disasters, the probability of major accidents caused by roof accidents is relatively small. However, due to the loss of control in control design decisions and implementation management (including unclear control of the range and changing rules of rich control strata, loss of control of support quality, etc.), major accidents such as large-scale bracket tilting or crushing still occur from time to time. Especially with the improvement of the equipment level of the fully mechanized mining face, high mining height is gradually being promoted and used in thick and extra-thick coal seam mining areas, the mining height is increased, the chance of large-span hanging roof hard rock layer appearing in the direct roof increases, the structure model of the mining area changes accordingly, and the mine pressure is more intense. Therefore, deepening the research on the control design and key technologies of roof control and related dynamic information foundation, solving the informatization, intelligence and visualization of roof control decision-making and implementation management, is still an urgent task for coal mine safety production. Based on the current situation of common roof accidents in China's coal mines, the causes and conditions of roof disaster accidents in the mining face were discussed. On the basis of studying the existing roof movement rules, a mining area roof structure model was established based on the overburden structure characteristics of the mining area. The working load and shrinkage determination methods of the bracket under the "given deformation" and "limited deformation" mechanical states were proposed and analyzed, the concepts of direct roof and basic roof under high mining height were revised, the design criteria and load calculation methods of the bracket control roof in the high mining height mining area were established, and the control theory and related information foundation were discussed in detail. The research results provide a basis for the prediction and control of roof disasters, especially the control design and bracket selection calculation of the roof in the high mining height mining area [29–32].

4.3.1 Overview of Roof Disaster Accidents in the Mining Face

1. Causes and Conditions of Roof Accidents in the Mining Face

The causes and conditions of roof fall (collapse) accidents during the excavation process of the mining face can usually be summarized as follows: unbalanced roof damage and movement; untimely support; insufficient support impedance or poor stability. Correctly predicting the location and time of roof instability and destruction movement, and setting up a strong and stable support in time before the roof instability movement, is the key to controlling accidents.

Roof accidents in the mining area can be divided into local roof fall accidents (occurring in the broken part of the roof) and large-area roof collapse accidents

4.3 Key Technologies for Roof Disaster Prevention and Control

(occurring at the moment of direct roof or basic roof pressure movement); according to the source of the accident, they can be divided into: roof accidents caused by direct roof (pressure) movement, including local roof fall of broken direct roof, large-area roof collapse accidents caused by direct roof movement crushing and pushing down the bracket; roof accidents caused by basic roof (pressure) movement, including basic roof pressure impact destruction and pushing down the bracket causing large-area roof collapse accidents. According to the characteristics of roof collapse movement and bracket instability destruction, it can be divided into crushing accidents, that is, when the direct roof and basic roof come to press, the bracket impedance is insufficient or the shrinkage is not enough to be crushed and destroyed (the support is bent or broken, etc.), causing roof collapse accidents; pushing down accidents, that is, when the direct roof and basic roof come to press, the bracket is pushed down due to insufficient stability, causing roof collapse accidents (Fig. 4.19).

2. The Relationship between Mining Stress and Roof Disaster Accidents

In the case of a thin-layered composite roof (which is prone to fracture under mining stress) or when the mining coal seam burial depth, mining height, and mining width (working face length) are large, the peak of the mining stress field penetrates deep into the coal wall (working face) ahead, that is, when the internal stress field appears, if the support is not timely, local roof fall accidents are likely to occur. The time and place of local roof fall accidents under the above conditions are determined by the law of pressure movement of the basic roof. Theory and practice have proven that

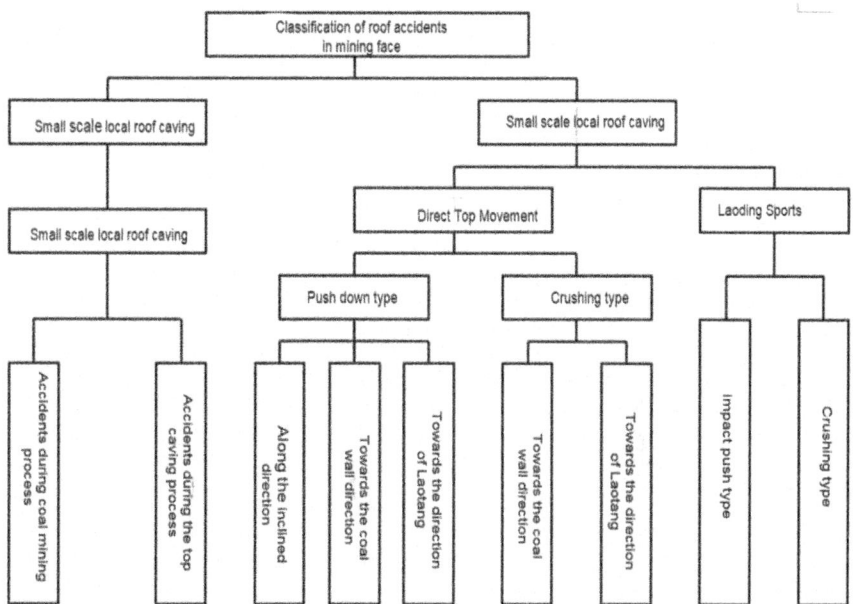

Fig. 4.19 Roof accident classification system diagram

most local roof fall accidents (collapse of roadway roof) occur during the pressure process of the basic roof. Especially when the basic roof is under pressure, the part of the composite roof that is forced to fracture. The main reason for the fracture of the composite roof at this part is: the basic roof fractures, the mining stress field (stress field) is clearly divided into two parts, there will be high-intensity pressure concentration near the roof fracture line; during the pressure rotation process of the basic roof fracture, the forced fracture rotation of the composite roof will transition from the original three-phase compression state to the single-phase compression state, and the compressive strength will drop sharply; the stress peak of the external stress field crushes the low-strength composite roof through pressure fracturing, realizing the transfer to the deep part of the coal wall. In summary, the time and place where local roof fall accidents may occur are regular. The occurrence of the law and the realization of the conditions are directly related to the movement of the basic roof rock beam and the relative stress of the mining stress near the roof fracture line, concentration and transfer.

Roof collapse accidents, such as roof crushing brackets, large-scale collapse accidents occur during the pressure process of direct roof and basic roof large-scale rock fracture. For the case where the basic roof pressure fracture penetrates deep into the coal wall ahead (with an internal stress field), the large-scale roof pressure crushing bracket collapse accident is realized under the following conditions: the bracket impedance cannot balance the force of the large-scale movement of the direct roof, and it is fractured during the roof settlement process, losing its support ability; the bracket impedance is sufficient to balance the sinking force of the direct roof, but the allowable compression amount cannot adapt to the requirements of the basic roof settlement (to the given deformation position state), under the joint action of the basic roof and the already forced fracture settlement of the direct roof, it is crushed or broken, losing its support ability, this is an important reason for the occurrence of crushing accidents on the working face using rigid brackets.

Under the condition of the "internal stress field", the characteristics of the roof fracture pressure movement process in the mining field. It is that the roof fracture penetrates deep into the coal wall (working face) ahead, and the large-scale roof settlement movement is realized by the coal body of the "internal stress field" as the support operation. Therefore, as long as before the basic roof movement that may produce dynamic pressure shock, try to maintain the "close" (minimum delamination) state of its lower rock layer (that is, the impedance of the bracket is to take the "limited deformation" working state for the movement of its lower rock layer) Under the premise, through the shrinkage of the bracket to adapt to the rock beam sinking to the bottom (contacting the old pond has fallen rock layer), the crushing type roof collapse accident can be avoided, this is exactly why under deep mining conditions, the bracket impedance of the mining field can often be much lower than that of the shallow mining field. Under the condition of the "internal stress field", the roof pressure crushing working face accident first starts from the crushing failure of the local section pillar of the mining field (start). If the impedance of the bracket in this section cannot be restored through timely supplementary support during the roof settlement process, and it cannot be supplemented from the support ability of the

4.3 Key Technologies for Roof Disaster Prevention and Control

adjacent section bracket. The roof settlement pressure destroys the bracket accident will quickly extend to the entire roof range, until the occurrence of a large-scale roof collapse accident. Therefore, correct support design, including the correct choice of support impedance and allowable shrinkage, is the key to controlling the occurrence and development of crushing type roof collapse accidents. Crushing type roof collapse accidents, when the roof is under large-scale pressure, start from the local section bracket dismantling damage (start), and quickly extend to the full range. The roof collapse accident where the bracket is crushed is always accompanied by the periodic fracture of the upper rock beam of the basic roof. It fully shows that under the condition of no "internal stress field" working face, especially when the roof is relatively hard, without high impedance brackets, to prevent the settlement of the lower rock beam and the direct roof, the upper high-strength rock beam movement will inevitably collapse the working face.

The collapse of the push-over type roof occurs during the roof pressure process, and the push-over power may come from the direct roof itself, or it may come from the impact and push of the basic roof fracture pressure. The conditions for the realization of the push-over type roof accident are as follows: ① There is a large delamination space between the direct roof and the basic roof boundary layer or between the layers of the composite structure direct roof. Thus losing the frictional resistance ability of layer movement. The conditions for realization are: the design resistance ability of the bracket (including the initial collapse force and working resistance) is insufficient; the quality of the bracket installation is not high, especially the initial support force is insufficient; the roof and floor are loose, the bracket breaks the roof and drills the bottom or because of the use of low roundness shoes (backboard) and other auxiliary support structures, resulting in the actual (effective) support (bearing) ability of the bracket is very low. ② The direct roof or part of its layers are in an isolated state with all around cut (cut off), the conditions for realization are: the old pond direct roof falls; the direct roof comes to press or under the basic roof pressure forced to crack along the coal wall (or deep into the front of the coal wall). ③ Faults and other structures cut off, or the recovery roadway roof breaking excavation cutting makes the direct roof cut off in the inclined direction (vertical advancement direction).

4.3.2 Mechanism of Roof Disaster Accident

The prediction of roof pressure is based on the correct understanding of the movement and destruction law of the overlying rock layer in the mining field. Among them: the law and mechanical process of the first fracture pressure of the "transfer rock beam" of the direct roof and the basic roof, as shown in Fig. 4.20.

When the rock beam advances to the first fracture step distance (C_0) in the mining field, it is in a fixed support state at both ends (Fig. 4.20a). The two moments $M_A = M_B = \frac{qC_0^2}{12}$ reach the limit value of the tensile stress at the end. At this time, the

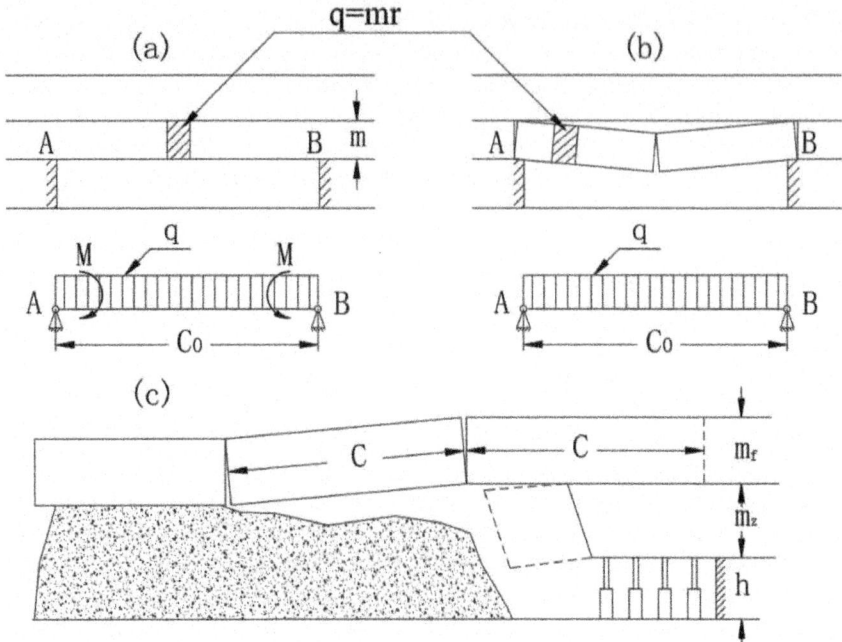

Fig. 4.20 Mechanical model of fracture pressure law

bending moment in the beam $M_0 = \frac{qC_0^2}{24} = \frac{M_A}{2}$. Only half of the end. Therefore, the fracture will start from the end that penetrates into the coal wall.

During the realization of the fracture at the end of the rock beam, the bending moment will gradually shift to the middle until the end of the beam is completely fractured into a simply supported state (Fig. 4.20b). At this time, $M_A = M_B = 0$, the bending moment in the beam M_0 reaches the maximum value. That is: $M_0 = \frac{qC_0^2}{8}$. Although, its value is far more than the bending moment value ($\frac{qC_0^2}{12}$) when the end fracture is realized. Therefore, after the end of the rock beam is fractured, the first pressure in the beam fracture mining field occurs.

When the mining field enters the normal advancement stage, the "direct roof" and "basic roof" each "transfer rock beam" is in the "cantilever beam" support state. Among them, the "direct roof" in a completely self-cantilever state, the fixed end bending moment $M_0 = \frac{qC_0^2}{2}$. Therefore, its maximum cantilever span C value will not exceed the first rock break step distance $\frac{1}{6}$, that is $C = \frac{1}{6}C_0$. Its fracture location depends on the resistance of the mining field bracket, especially the size of the initial support force may appear in the following two situations:

When $P_T \geq q_0(l_K + C)^2/l_K^2$ the direct roof is cut off behind the bracket;
$P_T < q_0(l_K + C)^2/l_K^2$ the direct roof may be cut off at the coal wall.

4.3 Key Technologies for Roof Disaster Prevention and Control

In the formula, P_T is the bracket strength (kN/m^2); l_K is the bracket control roof distance; C is the maximum suspension distance on the old pond side of the direct roof; q_0 is the bulk density of the direct roof (kN/m^2).

When the basic roof comes to press, the direct roof will be forced to cut off at the coal wall, and its pressure step distance is determined by the basic roof pressure. Therefore, correctly predicting the basic roof pressure is the key to roof control in the normal advancement stage.

Entering the normal advancement stage, each rock beam of the "basic roof" is in a cantilever state with one end touching the gangue, as shown in Fig. 4.20c. Studies have shown that its periodic fracture step distance C from the first fracture of the rock beam to the beginning of a small and large gradual transition to a constant equal to the first pressure step distance C_o of $\frac{1}{3}$ or so.

The periodic rock beam breakage of the basic roof starts from the end breakage. The rock beam realized by the end breakage will sink under the force of gravity, and the pressure from the mine roof will also begin.

In summary, under normal roof conditions, whether it is the first pressure stage of the roof rock beam or the periodic breakage pressure of the rock beam entering the normal advancement stage, it all starts from the breakage of the embedded end of the rock beam.

4.3.3 Basic Information for Roof Disaster Accident Prevention and Control

1. Key technology for roof disaster accident prevention and control

Based on the realization of the movement and pressure prediction of the mine roof, targeted and correct roof control design is carried out. The main work content of the roof control design decision includes: the establishment of the "structural mechanics model" centered on the range and support conditions of the roof to be controlled and the determination of related parameters; the relationship between the impedance capacity of the mine support and the position status (position) of the roof to be controlled, the determination of safety control criteria and related mechanical guarantee conditions, etc. The roof control design decision model (control design structural mechanics model) includes two parts: the range of rock layers to be controlled and the support conditions. The model structure and parameters are shown in Fig. 4.21.

(1) The range of rock layers to be controlled includes two parts: the "direct roof" that has collapsed in the old pond and the "basic roof" composed of the "transmission rock beam" affected by the movement.

The expression for the thickness of the direct roof is as follows:

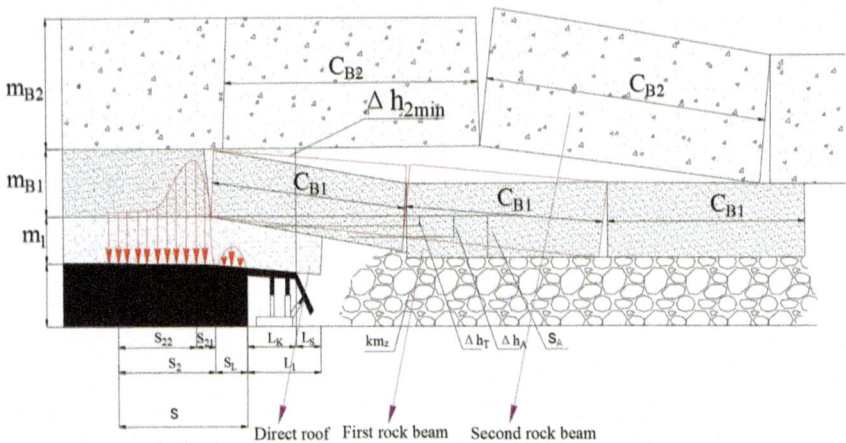

Fig. 4.21 Structural mechanics model of roof control design

$$M_Z = \sum_1^n M_i = \frac{h - S_A}{K_A - 1} \qquad (4.25)$$

In the formula, n is the number of rock layers that have collapsed in the old pond; M_i is the thickness of the collapsed rock layer, h is the mining height, K_A is the expansion coefficient of the collapsed rock layer, S_A is the settlement value at the confirmed waste place of the rock beam below the basic roof (always less than the old settlement value of the rock beam S_0).

Common methods for determining the thickness of the direct roof include the "measured inference method" inferred from the measured step distance C_0 and the corresponding mine roof subsidence Δh_0 using expression 4.26. And the "drilling columnar inference method" judged by the drilling columnar diagram of the overlying rock layer of the mine.

The inference procedure of the "measured inference method" is:

Step one: Measure the first pressure step distance of the rock beam below the basic roof C_0 and the corresponding control roof distance L_k subsidence of the mine roof h_0;

Step two: Connect formula 4.26 to calculate the settlement value at the waste place of the lower rock beam S_A;

$$S_A = \frac{C}{l_K} \Delta h_0 \qquad (4.26)$$

Step three: Use expression 4.25 to infer the thickness of the direct roof.

$$M_Z = \frac{h - S_A}{K_A - 1}$$

4.3 Key Technologies for Roof Disaster Prevention and Control

In the formula, the expansion coefficient K_A value represents the rock strength of each rock layer of the direct roof. The higher the rock strength, K_A the greater the value. Generally, it can be taken $K_A = 1.25-1.35$.

The "drilling columnar inference method" judges layer by layer from bottom to top according to the principle that the thickness of each rock layer of the direct roof is less than the free space height allowed for movement below it, that is:

$$M_Z = \sum_1^n M_i \qquad (4.27)$$

where

$$M_n \leq h - \sum_1^{n-1} M_i(K_A - 1)$$

$$M_{n+1} > h - \sum_1^n M_i(K_A - 1)$$

The "basic roof" is composed of "transmission rock beams" that significantly affect the appearance of mine pressure in the mine (including roof pressure appearances such as support load and subsidence, and support pressure appearances such as coal wall compression). Each "transmission rock beam" in the "basic roof" is composed of rock layers that move at the same time (almost at the same time), including the support layer and the follower layer. Judged by the following formula.

If adjacent rock layers move at the same time (forming the same rock beam), then:

$$E_S M_S^2 \geq (1.15-1.25)^4 E_C M_C^2 \qquad (4.28)$$

If adjacent rock layers move separately (forming "rock beams" separately), then:

$$E_S M_S^2 < (1.15-1.25)^2 E_C M_C^2 \qquad (4.29)$$

In the above formula, M_S and E_S represent the thickness and elastic modulus of the lower rock layer respectively; M_C and E_C represent the thickness and elastic modulus of the upper rock layer respectively.

Research and practice have shown that under general roof (overlying rock layer) conditions, the number of "transfer rock beams" that make up the "basic roof" does not exceed three. The total thickness is between 4–6 times the height. The composition of the typical roof (overlying rock layer) structure of a single rock beam and multiple rock beams and their mining field rock pressure manifestations are shown in Fig. 4.22a, b.

The pressure step distances of each transfer rock beam in the "basic roof" under its own weight are expressed by formulas 4.30 and 4.31:

The first rock break pressure step distance (C_0):

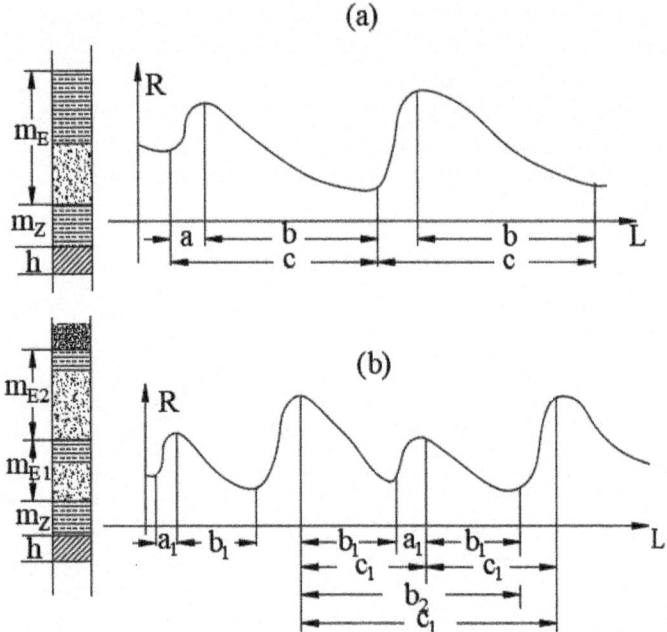

Fig. 4.22 Basic roof rock layer structure and mining field rock pressure manifestation

$$C_0 = \sqrt{\frac{2M_S^2[\sigma_S]}{(M_S + M_C)\gamma}} \qquad (4.30)$$

Periodic pressure step distance (C_i):

$$C_i = \frac{1}{2}C_{i-1} + \frac{1}{2}\sqrt{C_{i-1} + \frac{4M_S^2[\sigma_S]}{3(M_S + M_C)\gamma}} \qquad (4.31)$$

In the above formulas, M_S and M_C represent the thickness of the lower (support) rock layer and the upper (follower) rock layer of the rock beam respectively. $[\sigma_S]$ is the allowable tensile stress of the lower (support) rock layer. γ is the average bulk density of the rock beam. C_i and C_{i-1} represent the current periodic pressure step distance and the previous periodic pressure step distance associated with it.

If there is only one layer of rock moving at the same time (this layer), then the follower rock layer in the above formulas M_C is zero. This "rock beam" is composed of a single rock layer movement.

(2) The relationship between the resistance capacity of the mining field support and the state of the rock layer to be controlled—"transfer rock beam" position equation.

4.3 Key Technologies for Roof Disaster Prevention and Control

The resistance capacity and mechanical properties of the mining field support must ensure absolute control of the direct roof break pressure. That is, when the direct roof is cut off at the coal wall, the support must fully bear all its forces, which means that the support must ensure safe operation under the "given load" of the direct roof break pressure. The working resistance of the support that meets this requirement can be expressed by the following formula:

$$P_A = A = M_Z \cdot \gamma_Z \cdot f_Z \tag{4.32}$$

In formula 4.32, M_Z and γ_Z represent the thickness and average bulk density of the direct roof respectively; f_Z is the moment coefficient considering the position of the support combined force action point and the old pond hanging roof. When the distance from the support combined force action point to the coal wall (L_i), the control roof distance (L_k) and the hanging roof distance L_S are known. It can be calculated using the following formula:

$$f_Z = \frac{L_k}{2L}\left(1 + \frac{L_S}{L_K}\right)^2 \tag{4.33}$$

When the "basic roof" rock beam is under pressure, the direct roof will be forced to break and press. At this time, the mining field support can work under the working state of the rock beam old pond end (break point) settlement touch coal—"given deformation" condition. It can also work under the "limited deformation" condition that prevents the rock beam from sinking to the break point touch coal.

When the basic roof rock beam is under pressure. When the support works under the "given deformation" condition, the mining field roof subsidence (Δh_T) is "given" by the position state of the rock beam freely sinking to the old pond end break point touch coal, that is:

$$\Delta h_T = \Delta h_A \tag{4.34}$$

where

$$\Delta h_A = \frac{L_K \cdot S_A}{C_E} \tag{4.35}$$

$$S_A = h - m_Z(K_A - 1) \tag{4.36}$$

In the formula, Δh_A is the mining field roof subsidence under the "given deformation" (old pond break point can touch coal) condition of the mining field support; ΔS_A is the settlement value at the break point of the rock beam; L_k is the control roof distance of the support; C_E is the step distance for rock beam fracture; M_Z is the direct roof thickness for roof fall; K_A is the rock layer fracture coefficient for rock beam energy discharge; h is the mining height.

The resistance capacity of the support under the "given deformation" condition (P_T) can be arbitrarily selected between the following two limit values. Among them: the upper limit value of the support working under the "given deformation" condition cannot exceed the sum of the direct roof action force (A) and the force acting on the support from the sinking of the rock beam (K_A), that is:

$$P_{T\max} = A + K_A \tag{4.37}$$

where

$$A = m_Z \cdot \gamma_Z \cdot f_Z$$

$$K_A = \frac{m_E \cdot \gamma_E \cdot C_E}{K_T \cdot L_K}$$

In the formula, K_T is the proportion coefficient considering the weight of the rock beam borne by the support, which can be taken as $K_T = 2$ under general roof conditions.

The minimum resistance capacity of the support under the "given deformation" condition ($P_{T\min}$) cannot be lower than the direct roof action force, that is:

$$P_{T\min} = A = m_Z \cdot \gamma_Z \cdot f_Z \tag{4.38}$$

When the support is working under the "limited deformation" condition, the subsidence amount of the mining roof to be controlled (Δh_T) will be less than the subsidence amount of the mining roof when the rock beam can discharge energy at the fracture point (Δh_A), that is:

$$\Delta h_T < \Delta h_A$$

where

$$\Delta h_A = \frac{L_K S_A}{C_E} S_A = h - m_Z(K_A - 1)$$

The symbol definition in the formula is the same as before.

The necessary resistance capacity of the support under the "limited deformation" condition (P_T) can be expressed by the following "position equation" according to the known differences in rock beam structural parameters. Among them:

When the thickness of the rock beam M_E and the fracture step distance C_E are known:

$$P_T = A + K_A \frac{\Delta h_A}{\Delta h_T} \tag{4.39}$$

4.3 Key Technologies for Roof Disaster Prevention and Control

where

$$A = m_Z \cdot \gamma_Z \cdot f_Z \tag{4.40}$$

$$K_A = \frac{m_E \cdot \gamma_E \cdot C_E}{K_T \cdot L_K} \tag{4.41}$$

The meanings and calculation methods of the symbols in the formula are the same as before.

When the rock beam structural parameters M_E and C_E are unknown, the "position equation" of the rock beam can be established through actual measurement. Its expression is:

$$P_T = A + K_O \frac{\Delta h_O}{\Delta h_T} \tag{4.42}$$

where

$$A = m_Z \cdot \gamma_Z \cdot f_Z \tag{4.43}$$

$$K_O = P_O - A \tag{4.44}$$

In the formula, P_O and Δh_O are the roof pressure (support resistance) and the corresponding mining roof subsidence measured when the support is working under the "limited deformation" condition.

2. Basic information of roof disaster accident dynamics

The principle of roof control design is to strive for the lowest support strength and the smallest working resistance under the premise of effectively preventing roof fall accidents. The basic requirements include: preventing the direct roof from breaking, eliminating local roof fall accidents; preventing major roof accidents such as roof cutting and collapse; controlling the subsidence of the mining roof within the range allowed by the compression of the pillar (active pillar).

Practice has proved that during the pressure process of the basic roof fracture, the key to preventing the direct roof of the empty roof area (tunnel) from breaking and falling is to ensure that the resistance of the support can ensure the "internal stress field" range, (In the structural mechanics model shown in Fig. 4.23, the working face advances from A to B), the pressure borne by the coal wall reaches the lowest possible value (i.e. σ_{max}). That is, the resistance of the support should be able to ensure that the high-speed subsidence and rotation of the roof only occur after the fracture line enters the mining area (control roof).

The decomposition and control of major roof collapse accidents have proven that in the process of the direct roof and basic roof fracture pressing the preface and development, preventing (or minimizing as much as possible) the delamination

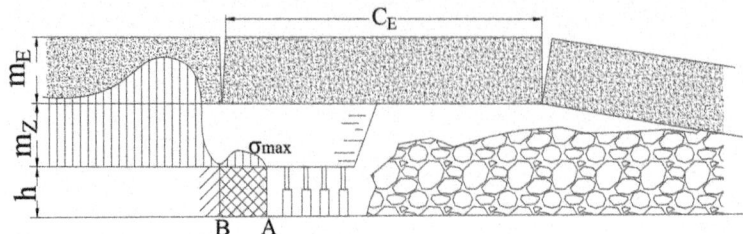

Fig. 4.23 Criterion and mechanical condition diagram

between the direct roof and the basic roof above the mining area (within the control area) and the delamination between the "lower rock beam" and "upper rock beam" of the basic roof is the key to eliminating the roof cutting collapse accident. The corresponding control criterion is that when the basic roof "upper rock beam" is pressed, the impedance of the support is sufficient to keep the "lower rock beam" working at the position shown in Fig. 4.24a, closely attached to the "upper rock beam". The required support strength (PT value) in its corresponding mechanical conditions can be obtained from the following "positional equation", namely:

$$P_T = A + K_A \frac{\Delta h_A}{\Delta h_T} \tag{4.45}$$

where

$$A = m_Z \cdot \gamma_Z \cdot f_Z$$

$$K_A = \frac{M_E \cdot \gamma_E \cdot C_E}{K_T \cdot \ell_K}$$

$$\Delta h_A = \frac{L_K S_A}{C_{E1}}$$

$$S_A = h - m_Z(K_A - 1)$$

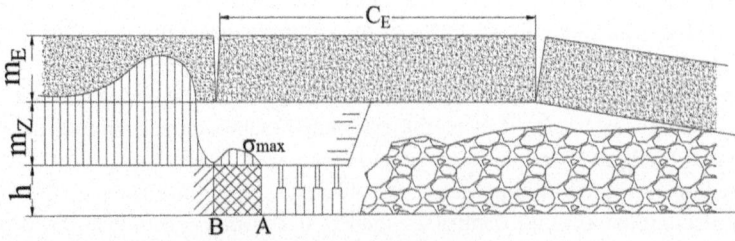

Fig. 4.24 Schematic diagram of design criteria and mechanical conditions

4.3 Key Technologies for Roof Disaster Prevention and Control

$$\Delta h_T = \Delta h_{\min}$$

In the formula, M_{E1} and C_{E1} are the thickness of the lower rock beam and the pressing step distance respectively; $\Delta h_{\min}$ is the minimum sinking amount of the mining roof before the upper rock beam (M_{E2}) is pressed, which can be approximately obtained by the following formula:

$$\Delta h_{\min} = \frac{L_K S_A}{2C_{E2}} \tag{4.46}$$

$$S_A = h - M_Z(K_A - 1) \tag{4.47}$$

Obviously, the support resistance under the above support strength conditions cannot limit the settlement of the upper rock beam. Therefore, when the upper rock beam is pressed, the support will be given deformation, and it will work under the conditions shown in Fig. 4.24b. The allowable shrinkage of the support (active column) under this condition must meet the following requirements. Otherwise, when the upper rock beam is pressed, a major accident of "pressing the support to death" will occur.

$$\varepsilon_{\min} \geq \Delta h_A - \sum \sigma_i \tag{4.48}$$

In the formula, $\varepsilon_{\min}$ is the required minimum allowable shrinkage of the support (active column); $\sum \sigma_i$ is the total compression of the support roof breaking, drilling bottom and "wearing shoes" "Wearing a hat" and other semi-automatic support structures, which can be considered as $\sum \sigma_i = 0$ during design; Δh_A is the sinking amount of the mining roof when the upper rock beam is pressed, which can be obtained by the following formula.

$$\Delta h_A = \frac{L_K S_A}{C_E}$$

$$S_A = h - m_Z(K_A - 1)$$

In the formula, C_E is the step distance of the forced fracture of the lower rock beam when the upper rock beam is pressed, which can be approximately replaced by the step distance of the free fracture of the lower rock beam; the meanings of other symbols are the same as before.

The information basis for controlling roof accidents in the mining face includes the following three aspects: ① Information on the range of rock layers to be controlled and their movement development rules. This mainly includes: the thickness of the direct roof and the step distance of the first collapse and periodic free collapse; the composition of the basic roof rock beam, the thickness of the related rock beam and the step distance of the first fracture and periodic fracture, etc. ② Information on the size distribution and development rules of the support pressure in the mining area. This mainly includes the location and time of the compression failure in front of the coal wall, also known as the "internal stress field", and the rules of expansion with the advancement of the mining area, etc. ③ The location and damage characteristics of faults and other structural damages, including the direction, drop, inclination angle (especially the angle between the fault surface and the working face) of the fault surface, and the structural features of major folding structures that penetrate the working face, etc.

The changes in mining height have altered the movement space of the overlying strata, the rotation angle of the rock mass, and the development of fractures. Based on the research and analysis of the mechanism of roof disaster accidents and prevention and control technology, in order to avoid the problem of roof control in the large mining height working face affecting the production capacity of the working face, the mining dynamics research group used the "transfer rock beam" theory to establish a large mining height mining field structural mechanics model, discussing the overlying strata structure and movement rules of the large mining height mining field, correcting the concepts of direct roof and basic roof under large mining height, and establishing the calculation method of support load. It is determined that in the "given deformation" state, the shrinkage of the support should be greater than the subsidence of the roof; in the "limited deformation" state, the impedance of the support is related to the position of the rock beam; as the mining height increases, the thickness of the direct roof may increase significantly, and the probability of large-span hanging roof hard rock layers appearing in the direct roof increases. The range of "transfer rock beam (basic roof)" that affects the appearance of mining pressure in the mining field is relatively reduced, and the height of the related "rock beam" from the mining field increases.

The range of rock layers that affect the appearance of mining pressure in the mining field is limited, known, and variable. The range of rock layers that have a significant impact on the appearance of mining pressure in the mining field is only a small part of the overlying rock layers, including "direct roof" and "basic roof". The relationship between the "support-surrounding rock" in the mining field includes the control method of the support for the direct roof and the control method for the basic roof beam. Under normal mining height conditions, the "given load" working method is adopted for the direct roof, and the basic roof adopts the "given" and "good" deformation working methods. Under the condition of large mining height, due to the possible appearance of large-span hanging roof structure in the direct roof. To ensure the safe and effective work of the support, the "given deformation" and "limited deformation" working methods should also be adopted for the direct roof. For the direct roof, in the "given deformation" working state, the position state of the direct roof when it is stable is determined by its strength and the support situation at both ends, that is, the shrinkage of the support meets the subsidence of the overlying rock layer, and the impedance is not enough to resist the subsidence of the direct roof, and can only reduce its movement speed within a certain range; in the "limited deformation" working state, the position state of the direct roof when it is stable is determined by the support situation of the support, that is, the support should be able

4.3 Key Technologies for Roof Disaster Prevention and Control

to completely impede the subsidence of the direct roof. For the basic roof, in the "given deformation" working state, the position state of the rock beam end touching the rock and the rock beam moving stably is determined by the strength of the rock beam and the support situation at both ends. During the entire movement process of the rock beam from the end fracture to the settlement to the final position, the support can only reduce the movement speed of the rock beam within a certain range, but cannot prevent the movement of the rock beam. In the "given deformation" working state, the relationship between the support force and the roof pressure during the entire process of rock beam movement is: Q roof > R support. In the "limited deformation" working state, the state of the rock beam end not touching the rock and entering stability (the subsidence of the mining field roof at the predetermined roof control distance when the rock beam movement is stable) is limited by the impedance of the mining field support, and in the "limited deformation" working state, the subsidence of the mining field roof limited by the resistance of the mining field support is less than the subsidence of the roof.

When mining thick coal seams with large mining height, due to the significant increase in the one-time mining thickness, the space of the mined-out area is relatively large, the collapse range of the direct roof and the fracture movement range of the basic roof will increase significantly, and the original direct roof collapse is not enough to fill the mined-out area, making the overlying rock layer of the mining field have a large rotation space, therefore, some rock layers originally belonging to the basic roof may be transformed into direct roof, the possibility (probability) of large-span hanging roof hard rock layers appearing in the direct roof increases, at the same time, the range of the basic roof (including the number and total thickness) that affects the appearance of mining pressure in the mining field is relatively reduced, the height of the related basic roof from the mining field increases, and the possibility of fracture movement impacting the dynamic pressure of the mining field significantly decreases. At this time, the structure model of the overlying rock layer in the mining field is shown in Fig. 4.25, and a large-span hanging roof appears in the direct roof.

The direct roof is generally defined as the sum of the collapsed rock layers in the mined-out area, and it cannot maintain the transfer force in the advancing direction, and the weight is fully borne by the support; the basic roof refers to the one that always maintains the transfer force in the advancing direction, and has a significant impact on the appearance of mining pressure in the mining field. Therefore, the previous

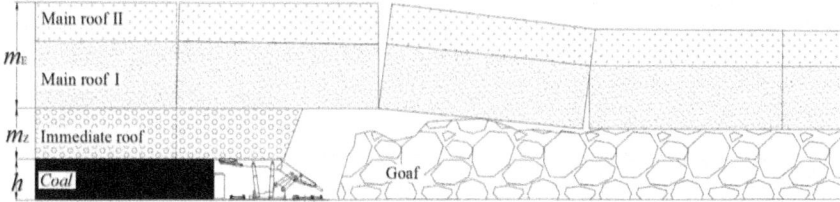

Fig. 4.25 "Single key layer" mining field structure. m_E—Basic roof thickness, m; m_Z—Direct roof thickness, m; h—Coal seam thickness, m

definition of the direct roof is no longer suitable for the situation of large mining height, and the definition of the direct roof under large mining height is revised to be the sum of the collapsed rock layers in the mined-out area, and it cannot maintain the transfer force in the advancing direction, and the weight is fully borne by the support or partially borne.

(1) Theoretical determination of the range of direct roof:

$$M_Z = \sum_{i=1}^{n} M_i, M_n \leq h - \sum_{i=1}^{n-1} M_i(K_A - 1), M_{n+1} \geq h - \sum_{i=1}^{n} M_i(K_A - 1)$$

In the formula, M_i is the thickness of the rock layer; K_A For the coefficient of gob-side stone swelling; h For mining thickness.

(2) Direct roof range measured:

$$M_Z = \frac{h - S_A}{K_A - 1} \quad (4.49)$$

In the formula, S_A is the size of the rock layer settlement.

(3) Basic roof range theoretically determined:

Adjacent rock layers move simultaneously (forming the same key layer), then

$$E_S M_S^2 \geq (1.15 - 1.25)^4 E_C M_C^2 \quad (4.50)$$

Adjacent rock layers move separately (forming different key layers), then

$$E_S M_S^2 < (1.15 - 1.25)^2 E_C M_C^2 \quad (4.51)$$

In the formula, E_S is the elastic modulus of the lower rock layer; E_C is the elastic modulus of the upper rock layer; M_S is the thickness of the lower rock layer; M_C is the thickness of the upper rock layer.

(4) Basic roof range measured:

If the overlying rock layer is a "single key layer" structure, during periodic fracturing, the support load value R and the advancement step distance L show a single periodic fluctuation state; if the overlying rock layer is a "double key layer" structure, during periodic fracturing, the support resistance curve shows a large and small periodic fluctuation state, small amplitude fluctuations represent the pressure appearance when the lower key layer fractures, large amplitude fluctuations represent the pressure appearance when the upper key layer fractures, as shown in Fig. 4.26.

The "support-surrounding rock" relationship in the mining area, including the control methods of the support for the direct roof and the basic roof, etc. Among them, the "given load" control method is adopted for the direct roof, and the "given

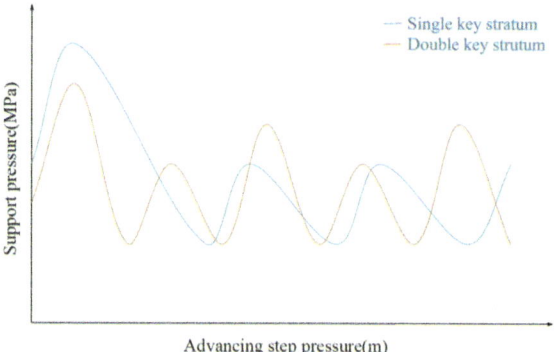

Fig. 4.26 Pressure manifestation rules of different key layer structures

deformation" and "limited deformation" control methods are adopted for the basic roof. Under the condition of large mining height, due to the possible large-span suspended roof structure of the direct roof, as shown in Fig. 4.27. To ensure the safe and effective operation of the support, the "given deformation" and "limited deformation" working methods should also be adopted for the direct roof.

(5) Control design of the "given deformation" working state of the direct roof:

The position state of the direct roof when it is stable is determined by its strength and the support at both ends, that is, the shrinkage of the support meets the subsidence of the direct roof, and the resistance of the support is not enough to resist the subsidence movement of the direct roof, and can only reduce its movement speed within a certain range. The structural model is shown in Fig. 4.28a.

The mining area is in a relatively balanced and stable state, take $f1 = 1, f2 = \frac{1}{2}$, according to the mechanical balance, establish the support balance equation

$$q_z = m_z \rho_z g f = m_{z1} \rho_{z1} g + \frac{m_{z2} \rho_{z2} g}{2} \tag{4.52}$$

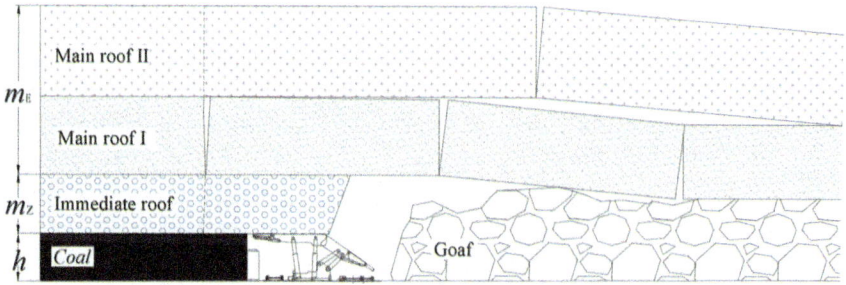

Fig. 4.27 "Double key layer" mining structure. M_{E1}—Thickness of the lower key layer of the basic roof, m; M_{E2}—Thickness of the upper key layer of the basic roof, m; C_{E1}—Pressure step distance of the lower key layer of the basic roof, m; C_{E2}—Pressure step distance of the upper key layer of the basic roof, m

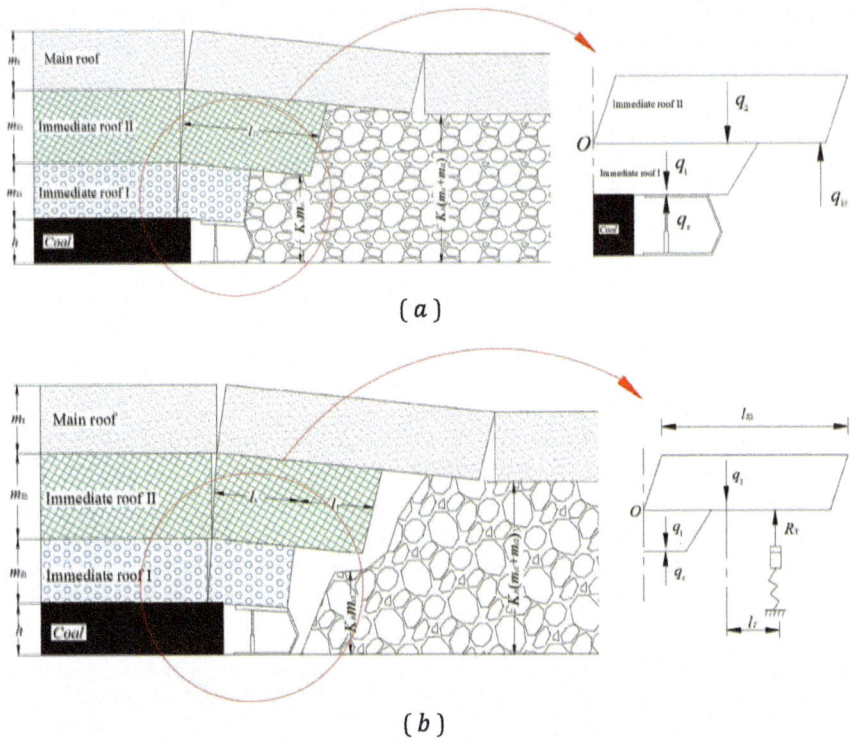

Fig. 4.28 Direct top "given deformation" and "limited deformation" mining field structure

Shrinkage of the support column

$$\varepsilon_{\text{shrinkage value}} = \Delta hA = \frac{h - m_{Z2}(K_A - 1)}{C_{Z2}} l_{Z2} \qquad (4.53)$$

In the formula, l_{Z2} is the control distance of the second layer of the direct roof, m; C_{Z2} is the pressure step distance of the direct roof, m; $K_A - 1$ is the coefficient of gob-side stone swelling; m_{Z1}, m_{Z2} are the first and second layers of the direct roof, m; ρ_{Z1}, ρ_{Z2} are the densities of the first and second layers of the direct roof, (kg/m³).

(6) Control design of the "limited deformation" working state of the direct roof:

The position state of the direct roof when it is stable is determined by the support of the support, that is, the support should be able to completely resist the subsidence of the direct roof. The structural model is shown in Fig. 4.28b.

The mining area is in a relatively balanced and stable state, according to the mechanical balance

4.3 Key Technologies for Roof Disaster Prevention and Control

$$R_T l_T = m_{Z1}\rho_{Z1} g l_{Z1} \frac{l_{Z1}}{2} + m_{Z2}\rho_{Z2} g l_{Z2} \frac{l_{Z2}}{2}$$
$$= \frac{1}{2}\left[m_{Z1}\rho_{Z1} g l_{Z1}^2 + m_{Z2}\rho_{Z2} g (l_k + l_f)^2\right] \quad (4.54)$$

Support strength

$$q_z = m_{Z1}\rho_{Z1} g + m_{Z2}\rho_{Z2} g + m_{Z2}\rho_{Z2} g \left[\left(\frac{l_f}{l_k}\right)^2 + 2\frac{l_f}{l_k}\right] \quad (4.55)$$

Know $l_f = 0$ when, $q_z = m_{Z1}\rho_{Z1} g + m_{Z2}\rho_{Z2} g = m_Z \rho_Z g$, $l_f = l_k$, $m_{Z1} = m_{Z2}$ when, $q_z = 5 m_{Z2} \rho_{Z2} g$.

(7) Basic top "given deformation" working state control design:

The position state of the key layer at the end of the key layer and the stable movement of the key layer is determined by the strength of the key layer and the support at both ends. During the entire movement process of the key layer from end fracture to settlement to the final position, the support can only reduce the movement speed of the key layer within a certain range, but cannot prevent the movement of the key layer. The mining field structure model is shown in Fig. 4.28a.

In the "given deformation" working state, the relationship between the support force and the roof pressure during the entire process of the key layer movement is

$$Q_{\text{roof}} > R_{\text{holder}} \quad (4.56)$$

At this time, during the entire process from the movement of the key layer to re-entering stability, it is impossible to establish a direct relationship equation between the support force and the roof pressure. But the support shrinkage can be obtained according to Fig. 4.28b

$$\varepsilon = \Delta h_A = \frac{h - m_Z(K_A - 1)}{C_E} l_K \quad (4.57)$$

where, C_E is the basic top pressure step distance, m.

(8) Basic top "limited deformation" working state control design:

The state of the key layer at the end of the key layer and the stable movement is limited by the impedance force of the mining field support. The structure model is shown in Fig. 4.29.

In the "limited deformation" working state, the shrinkage of the support and the sinking amount of the roof have the following relationship

$$\Delta h_{\text{holder}} < \Delta h_{\text{roof}} \quad (4.58)$$

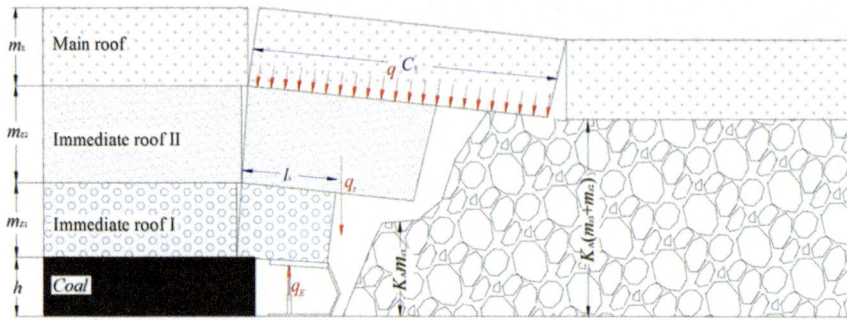

Fig. 4.29 Single key layer "limited deformation" mining field structure

Establish the mechanical relationship equation between the support resistance and the key layer position state that achieves balance, that is, the roof pressure received when the basic top sinking amount is Δh_i, including the basic top key layer force and the direct top force.

$$q_E = f(\Delta h_i) = q_Z + K \frac{\Delta h_A}{\Delta h_i} \qquad (4.59)$$

where, K is the key layer position constant, that is, the unit area key layer action when the roof sinking amount is Δh_A $K = \frac{m_E \rho_E g C_E}{K_T l_K}$.

In summary, the selection and design of the support should clarify the overlying rock layer structure, that is, determine whether it is "limited deformation" or "given deformation". When "limited deformation", the resistance is set, that is, the impedance force of the support should control the sinking value of the basic top down key layer to be sufficient to eliminate the possibility of impact when the upper hard key layer is pressed; "given deformation" is set to shrink, that is, the shrinkage can fully adapt to the maximum sinking amount of the mining field roof when the lower key layer touches the bottom (sinks to the bottom).

It is determined that the basic top key layer that has a significant impact on the appearance of mining pressure in the mining field can be divided into "single key layer" and "double key layer" structures; under the "double key layer" structure of the basic top, it is necessary to prevent the lower key layer from moving. The threats of cutting the top and over-limiting the active column shrinkage, and preventing the dynamic pressure impact on the mining field when the upper key layer is pressed. Based on the mechanical characteristics of the high mining field structure, the design guidelines for the high mining field support control top are established, as shown in Fig. 4.30.

(1) "Single key layer" structure

When the basic top is pressed, the support should be able to ensure its "limited deformation" on the basis of the direct top, and achieve the "given deformation" control requirements of the basic top;

4.3 Key Technologies for Roof Disaster Prevention and Control

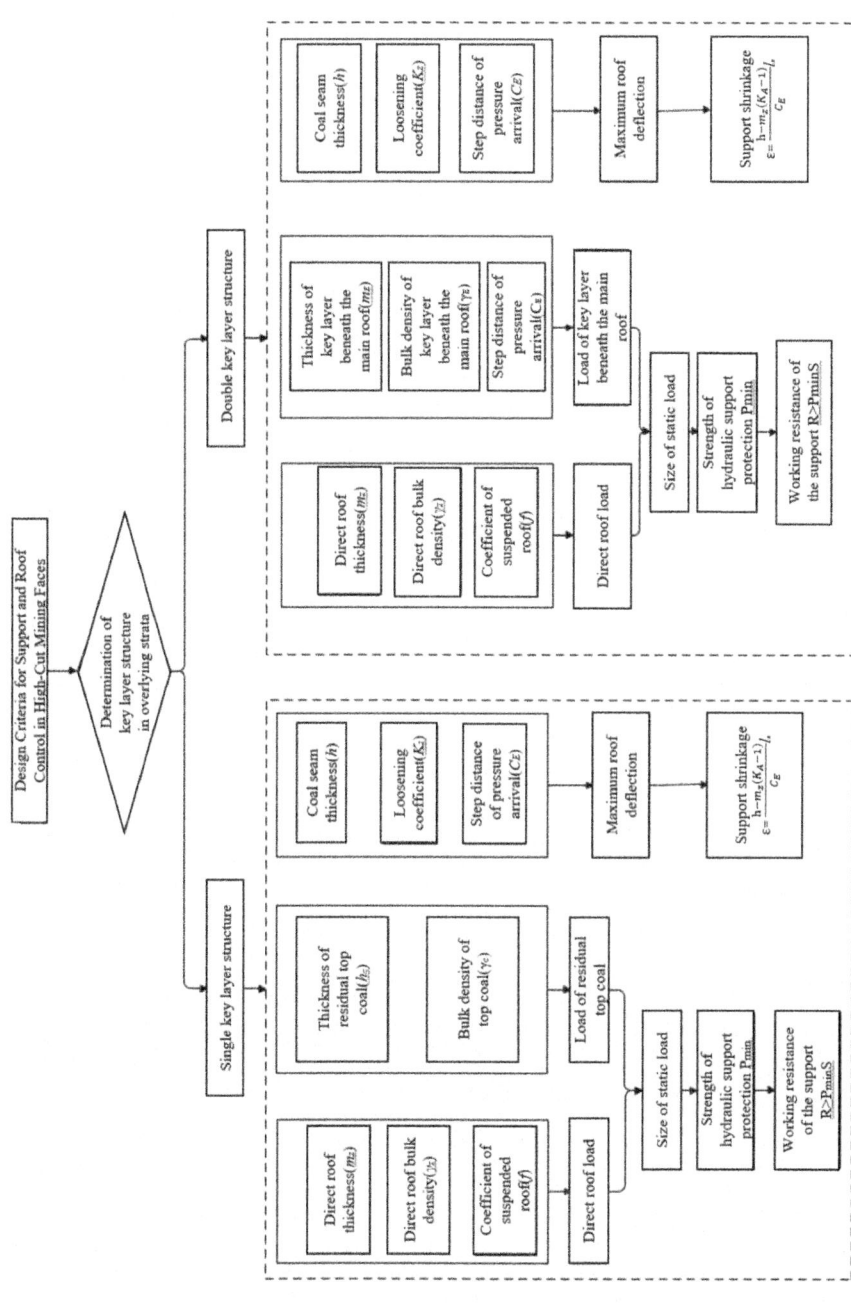

Fig. 4.30 Design criteria for bracket roof control in high mining height stope

As can be seen from Fig. 4.28, according to the mechanical balance, establish the support balance equation,

$$q_z = m_{coal}\rho_{coal}g + m_z\rho_z g$$

$$= m_{coal}\rho_{coal}g + m_{z1}\rho_{z1}g + m_{z2}\rho_{z2}g + m_{z2}\rho_{z2}g\left[\left(\frac{l_f}{l_k}\right)^2 + 2\frac{l_f}{l_k}\right] \quad (4.60)$$

where, m_{coal}, ρ_{coal} are the residual top coal thickness and density; m_{z1}, ρ_{z1}, m_{z2}, ρ_{z2} are the thickness and density of the first and second layers of the direct top; l_k, l_f are respectively the control roof distance and the suspended roof distance.

The bracket shrinkage is

$$\varepsilon = \frac{h - m_z(K_A - 1)}{C_E} l_K \quad (4.61)$$

In the formula, h is the coal seam thickness; K_A is the coefficient of gangue swelling in the mined-out area.

(2) "Double key layer" structure

When the basic roof comes under pressure, the bracket should ensure that on the basis of "limited deformation" of the lower key layer, the "given deformation" control requirements of the upper key layer are achieved;

When the basic roof comes under pressure for the first time, as can be seen from formula 4.59, the support strength of the bracket is

$$q_{E1} = q_z + K_A \frac{\Delta h_A}{\Delta h_i}$$

$$= m_{coal}\rho_{coal}g + m_z\rho_z g + \frac{m_{E1}C_{E1}\rho_{E1}g}{2l_k} \quad (4.62)$$

In the formula, C_{E1} is the first pressure step distance of the lower key layer of the basic roof.

When the basic roof comes under pressure periodically, as can be seen from formula 4.59, the support strength of the bracket is

$$q_{E2} = q_z + K_A \frac{\Delta h_A}{\Delta h_i}$$

$$= m_{coal}\rho_{coal}g + m_z\rho_z g + \frac{m_{E1}C_{O1}\rho_{E1}g}{2l_k} \quad (4.63)$$

In the formula, C_{O1} is the periodic pressure step distance of the lower key layer of the basic roof.

The shrinkage of the "double key layer" structure bracket is

$$\varepsilon = \frac{h - m_Z(K_A - 1)}{C_E} l_K \tag{4.64}$$

References

1. Pan J, Liu S, Ma W et al (2019) Intelligent prevention and control methods and development paths of deep impact ground pressure. Ind Min Autom 45(08):19–24
2. Wu S, Yang X, Guo L (2019) Thoughts and prospects on the overall planning of deep metal mines under high stress environment. J Coal 44(05):1432–1436
3. Wang J, Zhong L, Wang N (2017) Thoughts on underground mining technology and the development trend of underground mining. World Nonferrous Metals (18):17+19
4. Li T, Hao X (2009) Mechanism of dynamic disasters in deep mining and advanced identification. China University of Mining and Technology Press
5. Yuan L, Jiang Y, He X et al (2018) Research progress on precise identification and monitoring and early warning key technologies of typical dynamic disasters in coal mines. J Coal 43(02):306–318
6. Jiang L, Wei X (2015) Failure of steeply inclined slab-like rock mass tunnel under mining stress. J Liaoning Tech Univ (Nat Sci Ed) 34(01):15–20
7. Du X, Wang T (2017) Study on the connotation and usage range of impact ground pressure, rock burst and mine earthquake. Coal Chem Ind 40(3):1–4
8. Du X (2016) Study on the mechanism and prevention of impact ground pressure in thick hard coal strata. China University of Mining and Technology (Beijing), Beijing
9. Qi Q, Chen S, Wang H et al (2003) Relationship between impact ground pressure, rock burst, mine earthquake and their numerical simulation study. Chin J Rock Mech Eng (11):1852–1858
10. Xu G, Wang L, Li Y (2002) Study on the mechanism and criteria of rock burst. Rock Soil Mech 23(3):300–303
11. Xu L, Wang L, Li T (1999) Review of the current status of rock burst research at home and abroad. J Yangtze River Sci Res Inst 16(4):24.27
12. Qian Q (2014) Definition, mechanism, classification and quantitative prediction model of rock burst and impact ground pressure. Rock Soil Mech (1):1–6
13. Jiang Y, Pan Y, Jiang F et al (2014) Mechanism and prevention of impact ground pressure in coal mining in China. J Coal 39(2):205–213
14. He MC, Xie HP, Peng SP et al (2005) Research on rock mechanics in deep mining. J Rock Mech Eng 24(16):2803–2813
15. Qi Q, Chen S, Wang H et al (2003) Relationship between impact pressure, rock burst, mine earthquake and their numerical simulation study. Chin J Rock Mech Eng 22(11):1852–1858
16. Wang J, Li B, Zhang T (2010) Study on the occurrence range and mechanism of coal shot phenomenon during tunnel excavation. Ind Min Autom 36(07):39–41
17. Si G, Durucan S, Jamnikar S et al (2015) Seismic monitoring and analysis of excessive gas emissions in heterogeneous coal seams. Int J Coal Geol 149(49):41–54
18. Liu Y, Chi H, Zheng Q et al (2013) Review of microseismic technology and application research. Prog Geophys 28(4):1801–1808
19. Dou L, He X (2007) Research on grading prediction of coal mine impact ground pressure. J China Univ Min Technol 36(06):717–722

20. Pan J (2016) Impact initiation mechanism of impact ground pressure and its application. Coal Sci Res Inst
21. Pan JF, Mao DB, Lan H et al (2013) Current status and prospect of research on impact ground pressure prevention technology in China's coal mines. Coal Sci Technol 41(06):21–25
22. Lippmann H (1987) Mechanics of "Bumps" in coal mines: a discussion of violent deformations in the sides of roadways in coal seams. Appl Mech Rev 40(8):1033
23. Liu J, Zhai M, Guo X et al (2014) Theory and application of joint monitoring of vibration field and stress field for impact ground pressure. J Coal 39(2):353–363
24. Pan J (2019) Research on the initiation theory of coal mine impact ground pressure and its complete technology system. J Coal 44(01):173–182
25. Song ZQ, Jiang YJ, Yang ZF et al (2003) Research on the dynamic information basis for the prediction and control of major accidents in coal mines. Coal Industry Publishing House, Beijing
26. Wu Q (2014) Research progress, problems and prospects of mine water prevention and control and resource utilization in China. J Coal 39(05):795–805
27. Lv Y, Zhang Y (2013) Current status and development trend of domestic water-conducting fracture zone research. Coal Mine Mod (02):101–104
28. Gao Y, Shi L, Lou H et al (1999) Rules of floor water inrush and water inrush dominant surface. China University of Mining and Technology Press, Xuzhou
29. Wen Z, Tang J, Wang H (2011) Research on the mechanical model of large mining height mining field and the working state of the support. J Coal 36(S1):42–46
30. Wen Z, Zhao X, Yin L et al (2010) Research on the control model of large mining height roof and the reasonable bearing of the support. J Min Saf Eng 27(02):255–258
31. Sun B (2012) Research on the mechanism and control of large deformation of surrounding rock in mining roadway. Taiyuan University of Technology
32. Li R, Chen S (2017) Research on the prevention and control technology of roof disaster in the fully mechanized mining face with a large mining height of 7m shallow burial. Coal Eng 49(S2):9–13

Open Access This chapter is licensed under the terms of the Creative Commons Attribution-NonCommercial-NoDerivatives 4.0 International License (http://creativecommons.org/licenses/by-nc-nd/4.0/), which permits any noncommercial use, sharing, distribution and reproduction in any medium or format, as long as you give appropriate credit to the original author(s) and the source, provide a link to the Creative Commons license and indicate if you modified the licensed material. You do not have permission under this license to share adapted material derived from this chapter or parts of it.

The images or other third party material in this chapter are included in the chapter's Creative Commons license, unless indicated otherwise in a credit line to the material. If material is not included in the chapter's Creative Commons license and your intended use is not permitted by statutory regulation or exceeds the permitted use, you will need to obtain permission directly from the copyright holder.

Chapter 5
Control of Large Deformation of Surrounding Rock Based on Stress Gradient Theory

With the annual increase in the mining depth of coal mines and the exhaustion of coal resources in central and eastern China, deep mining is affected by high geothermal, high geostress, high osmotic pressure and mining disturbance. Some mines have transformed from shallow hard rock mines to deep soft rock mines. The Mesozoic coal-bearing strata in the western part have a late diagenetic age and poor cementation, and are loose and weak mudstone, sandy mudstone or muddy sandstone, which can easily modify when encountering water. This makes the problem of large deformation of deep roadway difficult to support increasingly prominent, especially the stability control of the mining roadway affected by dynamic pressure is more difficult [1, 2]. Ordinary anchor net rope cannot meet the requirements of roadway support, and there is a phenomenon of large deformation of the surrounding rock. If the support is improper, it will lead to fast sinking speed of the roadway roof, large sinking amount, and often accompanied by local roof fall; fast bottom drum speed, large total deformation; serious side spalling of the two sides of the roadway. This not only reduces the service life of the roadway, increases maintenance costs and economic costs, but also endangers the safety of underground personnel, is not conducive to the safe and efficient production of coal mines, and seriously affects the normal progress of mining work. Especially for the support control of broken surrounding rock under the influence of mining, it has always been a major problem in the mining industry. For a long time, the control of large deformation of roadway surrounding rock has always been a hot issue that people are committed to solving [3–8].

Its purpose is to ensure the normal use of the roadway and create necessary conditions for safe production in the mine.

For large deformation of the roadway, the existing mechanisms of large deformation in deep roadways are mostly due to the combined effects of high geostress (high self-weight stress caused by large mining depth, structural stress caused by fault groups, and superimposed stress caused by mining), weakening of coal and

Fig. 5.1 Broken surrounding rock in the mining area

rock properties, and support methods. The complex and abnormal high-stress environment is the fundamental cause of large deformation in deep well roadways [9, 10], the poor cementation degree of the surrounding rock, strong expansibility, low support strength, and the traditional anchor cable support is not coordinated with the deformation of the surrounding rock under the intense excavation disturbance, which cannot fully exert the bearing capacity of the surrounding rock itself, is the main reason for the destruction of weakly cemented expansive soft rock roadway. Under the conditions of deep well mining, the weakening of coal and rock properties after entering the deep part has obvious characteristics, that is, the surrounding rock shows continuous strong rheological characteristics, brittle coal and rock under low confining pressure can be transformed into ductility under high confining pressure, not only the deformation is large, but also has a significant "time effect" [11, 12]. Existing research divides large deformation of surrounding rock into three types: large deformation caused by large burial depth, large deformation caused by mining activities, and large deformation caused by low strength of the rock itself. Before the excavation of the roadway, the underground rock layer where it is located is in a natural equilibrium state. The excavation of the roadway or chamber destroys the original stress equilibrium state, causing the stress redistribution of the surrounding rock, changes in stress state and high stress concentration, resulting in displacement or rupture towards the excavated roadway or chamber. In the interaction process between the support structure and the surrounding rock, a load on the support is formed. When excavating a roadway or chamber, regardless of whether it is ultimately balanced or destroyed, the stress inside its surrounding rock will be redistributed, which is not subject to human will. This stress redistribution behavior is the process of self-organization stability of the roadway surrounding rock, and the large deformation of the roadway is also closely related to the stress situation of the deep rock mass, so in-depth study of mining stress and fully exerting the self-stability of the rock is the most economical and reliable method to achieve stability of rock underground engineering (Fig. 5.1).

5.1 Phenomenon of Large Deformation and Destruction of Surrounding Rock in the Mining Area

Whether the surrounding rock is broken is divided according to the integrity of the entire rock mass. Broken surrounding rock refers to those with three or more structural surfaces, even chaotic, mainly weathered or jointed and weathered fractures, greatly affected by tectonic stress or faults near the fault, fractures open, filled with fillings, and the integrity index of the entire rock mass is less than 0.55. At present, there are two methods for mining the overlying coal seam of the floor roadway in China: leaving a protective coal pillar to protect the roadway and cross-mining. Although leaving a protective coal pillar can avoid the severe mining impact on the floor roadway, the protective coal pillar is surrounded by mined-out areas, causing the area below the coal pillar to become a high-stress area. Under high stress, the surrounding rock of the floor roadway undergoes obvious rheological phenomena over time, which is not conducive to the long-term stability and maintenance of the roadway. If cross-mining is used for mining, stress increase areas will be generated in front of and laterally to the coal wall during the cross-mining process. As the working face continues to advance, the stability of the floor roadway will be affected by mining, leading to a decrease in the strength and stability of the surrounding rock of the roadway, and even destruction. The characteristics of broken surrounding rock are mainly poor stability, weak cohesion, discontinuous structure, and it is very easy to collapse during roadway excavation, especially when there is groundwater. If the original rock stress balance system is destroyed during the excavation of the roadway and effective support measures are not taken in time, it may cause the excavation roadway to collapse, directly threatening the lives of workers and equipment. Not only that, this situation will also prolong the construction period. In addition, if the tunnel is shallow, it is difficult to form a bearing arch during excavation, which will cause surface deformation and subsidence, affecting surface activities. According to statistics, 70–80% of the roadways in China are affected by mining, showing serious bottom drumming, large deformation of the surrounding rock and difficult to control. The maintenance of the roadway affected by mining has severely restricted the intensification of coal mine production [13–16]. Only by understanding the characteristics of broken surrounding rock and finding a reasonable and effective support method can we ensure the normal progress of tunnel construction and production activities and ensure the safety of construction workers [17].

5.1.1 Characteristics and Causes of Large Deformation and Damage of Surrounding Rock in Mining Area

The intact roadway surrounding rock medium can be approximated as a linear elastic body or an ideal elastoplastic body, and the deformation of the surrounding rock is small deformation. The key issue of this small deformation rock engineering is strength instability. However, the deformation of broken surrounding rock is essentially characterized by significant non-linearity, non-smoothness, and irreversible plastic deformation. The core issue of broken surrounding rock roadway is large deformation instability. Generally, the deformation and damage of broken surrounding rock have the following characteristics [18]: Various forms. The deformation and damage modes generally include sinking and collapse of the top and bottom plates, slabbing and bottom drumming, etc., and the surrounding rock shows strong overall convergence and damage phenomena. The deformation and damage modes are both structure-controlled and stress-controlled, with the majority being stress-controlled. Long time. After the roadway is constructed, the stress of the surrounding rock is readjusted to reach a new stress equilibrium state. Then, because the broken surrounding rock does not have enough strength and has strong rheological properties, the time for stress readjustment becomes longer, and the duration of deformation and damage is also extended. The deformation of the broken surrounding rock is large and the deformation speed is fast: the sinking amount of the tunnel roof is large, most of which is between 200–500 mm, the deformation of the two sides of the tunnel is serious, and the displacement of a single side is between 200–800 mm, accompanied by a strong drumming phenomenon at the bottom. The initial convergence speed of the broken surrounding rock reaches 30 mm/d. After using the conventional spray anchor support, the convergence speed of the surrounding rock can still reach more than 20 mm/d, and its deformation convergence speed decreases slowly.

The range is large. Under the action of large original rock stress, the broken surrounding rock is damaged more extensively than the intact surrounding rock due to insufficient strength to bear, especially when the support measures are improperly taken, the range is even larger, and the maximum can even reach 2–5 times the radius of the tunnel.

The positions are different. In different parts around the tunnel, the degree of deformation and damage is different, which reflects the intensity of the geostress where the weak broken surrounding rock is located varies with direction, and the rock mass has strong anisotropy. The difference in deformation and damage in direction often leads to uneven stress on the support structure, resulting in a large bending moment in the support structure, which is very unfavorable for the stability of the support structure.

The pressure comes quickly. The deformation convergence speed of the broken surrounding rock is high. In a short time, the surrounding rock comes into contact with the support structure and generates pressure. After the surrounding rock and the support structure interact, the deformation and damage of the surrounding rock does not stop immediately, but continues. This is because the surrounding rock has rheological properties. In the rheological process of the surrounding rock, the strength of the surrounding rock decreases, so the mine pressure increases with time.

Based on a comprehensive analysis of key coal mines at home and abroad, the main reasons for the large deformation and damage of the broken surrounding rock in the mining area are summarized as follows [19–21]:

(1) Dynamic pressure is an important reason for the deformation and damage of the tunnel floor.

When the tunnel is in a static pressure state, the vertical principal stress it receives σ_y is:

$$\sigma_y = \gamma H \tag{5.1}$$

In the formula, γ is the bulk density of the overlying rock layer of the tunnel, N/m^3; H is the depth of the tunnel floor, m.

When the overlying coal seam is mined, the weight of the overlying rock layer of the mined-out area will be transferred to the coal and rock mass around the mined-out area, forming a support pressure belt and causing stress concentration. The surrounding rock of the tunnel floor will be damaged under the influence of high mining pressure and cannot maintain stability.

(2) The support form is unreasonable.

The use of reasonable support technology is also an effective means to avoid deformation and damage of broken surrounding rock. After the tunnel is excavated, the surrounding rock changes from a three-way stress state to a two-way stress state. The surrounding rock moves towards the empty face. When the deformation does not exceed the elastic deformation range of the surrounding rock, it can be unsupported, otherwise it must be supported. Untimely support or insufficient support strength may cause the weak structure in the weak structure of the tunnel surrounding rock to gradually damage, destroy and even cause serious deformation instability phenomena such as drumming at the bottom and sides of the tunnel.

The failure of the support system and the deterioration of the stable state of the surrounding rock are mainly due to the following reasons [22], firstly, the strength of the support body is insufficient, the support force provided is limited, and under the influence of mining, the stress concentration degree is very high due to the high mining stress, and the support strength is not enough to resist the high mining stress. Secondly, the support and the deformation stiffness of the surrounding rock are not coupled. Under low confining pressure, the tunnel damage mode is mostly brittle failure, while under high stress, the surrounding rock often shows strong creep [23]. The deformation space of the surrounding rock in the tunnel under the support of the steel mesh shell anchor spray + arch shed is limited, and it cannot release enough deformation energy. Energy accumulation is prone to occur in weak parts of the support, leading to the destruction of the surrounding rock. Third, the use of anchor cables and other supports did not fully mobilize the bearing capacity of the deep surrounding rock. The original support method could only control the deformation of the shallow surrounding rock of the tunnel, and could not make the deep rock effectively bear the load of the shallow surrounding rock, achieving the effect of improving the stress state and bearing capacity of the surrounding rock, eventually leading to the gradual destruction of the surrounding rock from shallow to deep [24]. Fourth, due to difficulties in tunnel construction and maintenance, there are no bottom plate control measures, and the tunnel support structure as a whole, the deformation of the bottom plate is closely related to the stability of the top plate and both sides. The lack of support for the bottom plate becomes the weak link of the entire tunnel support, eventually leading to the overall destruction of the tunnel surrounding rock.

(3) The surrounding rock itself has poor properties. The properties of the rock mass determine the bearing capacity of the rock mass, mainly including the mineral composition of the rock mass, the type of cementation, the structure of the rock mass, fillers, the structural state, and the degree of weathering, etc.

The mineral composition of the rock is the main reason affecting its physical properties. For example, clay rock and shale, their own compressive strength and shear strength are low, when the rock mass contains a large amount of clay or other expansive minerals, it will show serious softening or modification after soaking in water, its strength becomes even lower, if such rock mass is under high geostress or other external load conditions, its deformation and destruction are more intense.

The physical properties of the rock itself are not only affected by its mineral composition but also by the type of cementation of its internal particles. Some clay rocks and most sedimentary rocks are connected by the cementing material between their internal particles. When such rocks are subjected to external loads, their physical properties are closely related to the cementing material and cementation type of their internal particles. The mineral composition and cementation type of the rock are the intrinsic reasons affecting its physical properties, while its structural state is the external reason affecting its properties. During the process of rock formation, due to the great influence of tectonic movement or other external environments on its formation, it forms a complex internal structure. Various structural states formed inside the rock mass will reduce the physical and mechanical properties of the rock, and the deformation and destruction of the rock often start from the fracture structure.

Generally speaking, the higher the hardness of the mineral, the higher the strength of the rock. Structural surfaces usually weaken the strength and self-stability of the rock. If the structural surface is filled with weak components such as mud, accompanied by rock weathering phenomena, the strength of the rock mass is even lower.

(4) Geostress. The original rock stress is the most fundamental cause of the deformation and destruction of the broken surrounding rock. Construction excavation breaks the original stress balance of the surrounding rock, the stress is readjusted, causing some places in the construction tunnel surrounding rock to show stress concentration. The original rock stress can be divided into self-weight stress and horizontal tectonic stress. The size of the self-weight stress only depends on the overlying rock layer and the depth of the tunnel. The deeper the burial, the greater the mass of the overlying rock layer, and the gravity acting on its support structure also increases accordingly. Generally, the horizontal tectonic stress is closely related to the activity and complexity of geological tectonic movement. During the process of geological tectonic movement, the rock mass

5.1 Phenomenon of Large Deformation and Destruction of Surrounding … 205

deforms to a certain extent in both horizontal and vertical directions under the action of external forces, forming different distributions and different forms of geological structures. If its support structure does not have enough force to bear its self-weight stress and horizontal tectonic stress, the support structure will become unstable and be destroyed, and the surrounding rock will also deform and be destroyed. Geostress is a necessary condition for the destruction of the surrounding rock, and to a certain extent determines the mechanical properties of the rock mass, such as the strength and elastic limit of the rock mass under three-way stress are significantly higher than under two-way stress. Geostress can also cause the rock mass to transform between brittleness and plasticity, such as the rock mass that appears as hard rock during shallow mining may also show large deformation, plastic flow and other characteristics of soft rock under high stress during deep mining.

(5) Other factors influence.

Groundwater is mainly divided into dynamic water and static water. Dynamic water has dynamic water pressure, and static water has static water pressure. The impact of dynamic water pressure on the physical properties of rocks is mainly manifested in the kinetic energy generated by the water flow causing the rock mass to displace and continuously eroding the cementing material between the rock mass, causing the rock mass to break and become unstable. The impact of static water pressure is manifested in the static water pressure reducing the friction between the fractured rock blocks, leading to the collapse and collapse of the surrounding rock. In addition, the chemical components in groundwater can also cause erosion, softening, and dissolution of the rock mass, leading to deformation and destruction of the rock mass, especially in muddy surrounding rock, groundwater not only makes the surrounding rock muddy but also causes its volume to expand, becoming a deformation and destruction prone point.

The location and cross-sectional shape of the roadway determine the stress state of the surrounding rock of the roadway. The stability of the roadway is best when the direction of the roadway is parallel to the direction of the maximum principal stress. The smoother the transition of the curves in each section of the roadway, the better the stress state of the roadway, and the more conducive to the stability of the roadway.

External factors affecting the deformation and destruction of broken surrounding rock also include tunnel construction technology, tunnel type and size, and the support technology used. There are various construction techniques for tunnels, and the impact on the deformation and destruction of the surrounding rock is also different. Common tunnel shapes include elliptical, circular, and circular arch shapes. The reason why most tunnels choose the above shapes is because these shapes can better disperse the original rock stress than rectangular or trapezoidal cross-sections, avoiding stress concentration at a certain point or surface and causing tunnel instability and destruction. The size of the tunnel is not the bigger the better, and the appropriate engineering size should be chosen according to different engineering realities.

In general, there are many factors affecting the deformation and destruction of broken surrounding rock, and the impact is either large or small but should not be ignored. During construction, it is necessary to make various factors restrict each other in order to achieve the purpose of tunnel stability.

5.1.2 Mechanical Characteristics of Large Deformation of Stope Surrounding Rock

The stability of the surrounding rock depends on the degree of fragmentation of the surrounding rock, which reflects the impact of the surrounding rock by tectonic movement. The stability of the tunnel is closely related to the stability of the surrounding rock. Therefore, the degree of fragmentation of the surrounding rock plays a leading role in the stability of the tunnel. The higher the degree of fragmentation, the easier it is for the surrounding rock to become unstable after the tunnel is excavated, the worse the stability, and the more difficult the later support. Understanding the destruction mechanism of broken surrounding rock can find ways to support such tunnels.

At the beginning of the last century, R. Felmer first proposed the elastic–plastic analysis method of surrounding rock deformation, and then H. Kastnerls revised it based on R. Felmer's research [25], this theory played an important role in tunnel support for a certain period of time. But the shortcomings of both are that they assume that after the surrounding rock of the tunnel is damaged, the strength of the surrounding rock continues to maintain the original strength value, causing the theoretical calculation value of the surrounding rock plastic zone to be small, and the tunnel support designed according to this theory is difficult to support the actual surrounding rock load, causing the support structure to be damaged. With continuous research on soft rock tunnels, some scholars have proposed that the surrounding rock is assumed to be an ideal brittle-plastic body. The most representative is the Alery strain softening theory calculation formula, which proposes that the strength value will drop sharply to a certain residual value after the rock mass is damaged. The plastic zone of the surrounding rock calculated by this theory is too large, and the result is that the support stiffness designed by this is too large, causing a certain waste. After a large number of tests, it is known that the strength value of the surrounding rock damage is gradually reduced to a certain value, showing obvious rheological and creep characteristics.

The destruction of soft rock belongs to the category of solid mechanics research, and the development of related theories also provides a new interpretation method for the analysis of the destruction mechanism of soft rock tunnels. Among them, the most prominent is the damage mechanics theory. Since solid mechanics mainly studies the mechanical properties of continuous media, many scholars have established related continuous media models and theories from different angles, greatly enriching the theoretical support of damage mechanics in the analysis of the destruction mechanism of soft rock tunnels.

Fig. 5.2 The influence of vertical stress on point M in the infinite plane [22]

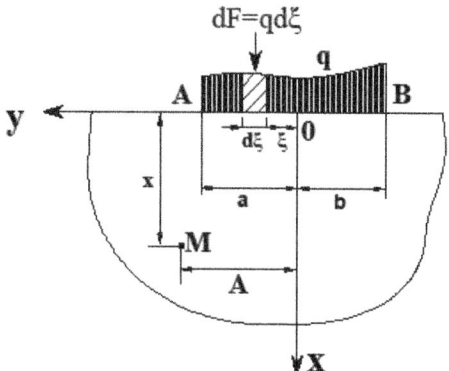

Since the nineteenth century, the theory of surrounding rock pressure has made great progress, from classical pressure theory and loose body pressure theory, to the currently widely used elastic mechanics theory and plastic mechanics theory. According to the elastic mechanics theory, when a semi-plane body is subjected to vertical distribution force on the boundary, as shown in Fig. 5.2, the distribution force can be taken at a small length ξ at the AB section distance from the coordinate origin O $d\xi$, the force received $dF = qd\xi$ is regarded as a small concentrated force, and the stress formula of the superposition of concentrated stress is obtained at any point M below, that is, formula 5.2.

$$\begin{cases} \sigma_y = -\dfrac{2}{\pi} \int_{-b}^{a} \dfrac{qx^3 d\xi}{[x^2+(y-\xi)^2]^2} \\ \sigma_x = -\dfrac{2}{\pi} \int_{-b}^{a} \dfrac{qx(y-\xi)^2 d\xi}{[x^2+(y-\xi)^2]^2} \\ \tau_{xy} = -\dfrac{2}{\pi} \int_{-b}^{a} \dfrac{qx^2(y-\xi) d\xi}{[x^2+(y-\xi)^2]^2} \end{cases} \quad (5.2)$$

Soviet scholar M. M. Protodyakonov, through studying the stress distribution of the rock mass above the underground ketone chamber, believes that after the surrounding rock stress is redistributed, a curved pressure arch is formed above the chamber, thus proposing the Protodyakonov theory [26–28], the calculation formula is:

$$q = \frac{\frac{B}{2} + h\tan\left(45° - \frac{\varphi}{2}\right)}{f} \gamma \quad (5.3)$$

where, q is the vertical average stress; γ is the weight of the surrounding rock; B is the tunnel width; h is the tunnel height; f is the hardness coefficient, φ is the internal friction angle of the surrounding rock. Through years of practical verification, the Prusik theory based on pressure arches can be well applied to deep tunnels in loose and broken geology.

Rock is a naturally formed brittle material. Due to long-term geological movements and complex geological environments, the rock contains rich cracks, holes, and defects, forming non-homogeneous, non-continuous, anisotropic, and strong discrete dissipative structural features. As people's understanding of the microscopic properties of rock materials continues to deepen, it is gradually found that the rock strength theory established by the classical continuum method cannot explain the randomness and discreteness of rock strength. Rock is a complex particle system formed by a large number of discrete particles through bonding. In fact, the failure process of rock is the bonding failure process between particles. Under the action of external forces, the binder between the particles inside the rock mainly bears tension, shear, pressure, and various forces. Under the action of external forces, the bond (also the bonding position) is broken to form micro-cracks. The aggregation of cracks forms a macroscopic failure surface. In fact, the failure of rock is the macroscopic manifestation of internal bonding failure accumulation.

Existing research shows that there are generally three types of formation mechanisms for large deformation of roadway surrounding rock: one is deep mining and deep underground engineering, due to the large depth of the project, resulting in a large rock self-stress field; the second is caused by underground mining activities, resulting in a stress field superposition effect; the third is that the rock itself has low strength and obvious deformation occurs under low stress conditions, which is a large deformation of expansive soft rock. For deep tunnels, their stress environment is quite different from that of shallow tunnels. The surrounding rock of deep tunnels has the characteristics of severe deformation and difficult control under high stress; the backfilling tunnel affected by mining, because the surrounding stress field has caused a disturbance once during excavation, and the superposition of mining stress during the working face backfilling, the tunnel is prone to large deformation.

Kang and others [29] took a deep roadway along the goaf in Huainan as the background, used numerical simulation to analyze the deformation of the roadway surrounding rock, and found that due to the large deep ground stress, the goaf along the roadway was more strongly affected by mining, the plastic zone of the coal body continued to expand, the coal gangue expanded and bulged, the roof subsidence and rotation continued to increase, and the bottom bulge was serious, as shown in Fig. 5.3; Fang and others [30, 31] studied the deformation characteristics of deep well broken surrounding rock roadway in Xuehu Coal Mine, and found that the two sides of the roadway moved closer, the deformation rate was large, the bottom of the roadway bulged seriously, and it lagged behind the excavation for half a month, and it was still difficult to self-stabilize after multiple repairs, as shown in Fig. 5.4.

The deformation and failure of the large deformation roadway surrounding rock is essentially caused by the formation and development of the plastic zone of the surrounding rock. The different geometric shapes and ranges of the plastic zone determine the failure mode and degree of the surrounding rock. Zhao and others [32, 33] According to the theory of elastoplastic mechanics, the boundary equation of the plastic zone of the surrounding rock of the circular roadway under the condition of non-uniform stress field is derived, and the irregular distribution of the butterfly-shaped plastic zone and the mechanical mechanism of the formation of the plastic

5.1 Phenomenon of Large Deformation and Destruction of Surrounding ... 209

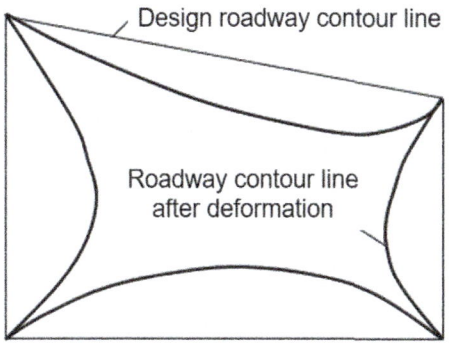

Fig. 5.3 Characteristics of large deformation of the roadway

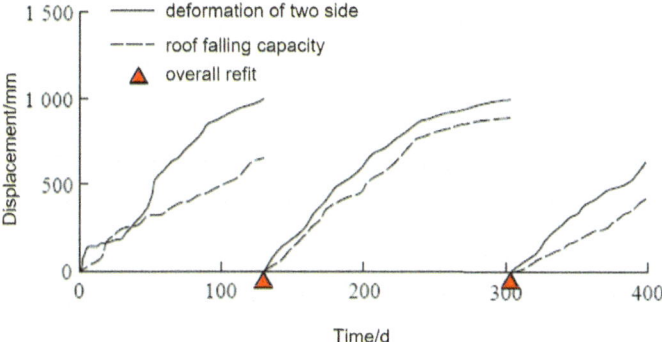

Fig. 5.4 Change curve of deep mine roadway surrounding rock

zone of the surrounding rock of the roadway are theoretically discovered. In order to directly use the method of elastic mechanics for analysis, the mechanical model of the mining roadway is simplified as follows: first, because the depth of the roadway is generally much larger than the radius of the roadway, the stress field can be regarded as a uniform load, and because the length of the roadway is generally large, it can be treated as a plane strain problem. Secondly, the strength of the surrounding rock and the non-uniformity and discontinuity of the rock mass are not considered first.

Instead, it is regarded as an isotropic homogeneous medium. The analysis of the circular roadway affected by the two-way unequal pressure stress field is shown in Fig. 5.5. Under the condition of the two-way unequal pressure stress field, according to the theory of elastoplastic mechanics, the boundary equation of the plastic zone of the surrounding rock of the circular roadway under the condition of non-uniform stress field is obtained, that is:

$$f\left(\frac{R_0}{r}\right) = K_1 \cdot \left(\frac{R_0}{r}\right)^8 + K_2 \cdot \left(\frac{R_0}{r}\right)^6 + K_3 \cdot \left(\frac{R_0}{r}\right)^4 + K_4 \cdot \left(\frac{R_0}{r}\right)^2 + K_5 = 0$$

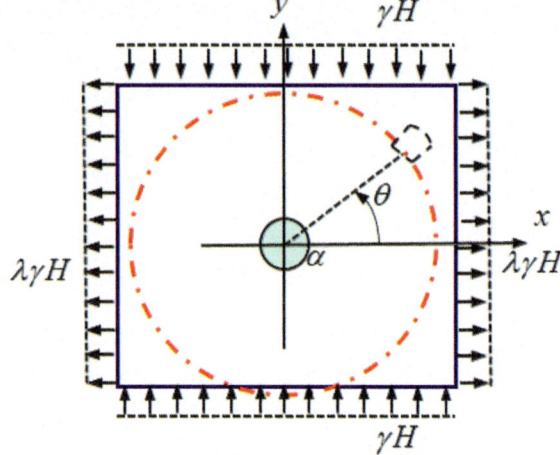

Fig. 5.5 Force model of unequal pressure circular roadway

where, R_0 is the radius of the circular roadway; r is the depth of the plastic zone at the corresponding θ angle; when the depth of the roadway, the bulk density of the surrounding rock of the roadway γ, the lateral pressure coefficient λ, the cohesion of the surrounding rock C and the internal friction angle φ are all given, the plastic boundary of the roadway can be calculated;

$$K_1 = 9(1-\lambda)^2;$$

$$K_2 = -12(1-\lambda)^2 - 6\left(1-\lambda^2\right)\cos 2\theta;$$

$$K_3 = 2(1-\lambda)^2\left[\cos^2 2\theta\left(5 + 2\sin^2\varphi\right) - \sin^2 2\theta\right] + (1+\lambda)^2 + 4\left(1-\lambda^2\right)\cos 2\theta;$$

$$K_4 = -4(1-\lambda)^2\cos 4\theta - 2\left(1-\lambda^2\right)\cos 2\theta\left(1 + 2\sin^2\varphi\right) + \frac{4}{\gamma H}(1-\lambda)\cos 2\theta \sin 2\varphi C;$$

$$K_5 = (1-\lambda)^2 - \sin^2\varphi\left(1 + \lambda + \frac{2C}{\gamma H}\frac{\cos\varphi}{\sin\varphi}\right)^2;$$

The ratio of the two-way load of the roadway (lateral pressure coefficient) has a significant impact on the geometric size and distribution shape of the plastic zone of the surrounding rock of the roadway. The plastic zone of the surrounding rock of the roadway in different orientations is sensitive to the change of the lateral pressure coefficient, resulting in an irregular "butterfly" plastic zone in the process of increasing the maximum ratio of the two-way load. This plastic zone has a large non-uniform coefficient, which is not conducive to the stability of the surrounding rock of the roadway. The larger the maximum ratio of the two-way load, the more likely the surrounding rock of the roadway is to have an irregular "butterfly" plastic zone, and the greater the maximum depth of the plastic zone of the surrounding rock of the roadway.

5.1 Phenomenon of Large Deformation and Destruction of Surrounding ... 211

The size of the load has an important impact on the geometric size of the plastic zone of the surrounding rock of the roadway. Under the condition of two-way load, the range of the plastic zone of the surrounding rock in the direction of larger load is smaller, while the range of the plastic zone produced in the direction of smaller load is larger. When the lateral pressure coefficient is constant, the size of the load has a greater impact on the size of the plastic zone of the surrounding rock, but has a smaller impact on the shape of the plastic zone, basically not changing the shape of the plastic zone of the surrounding rock.

The influence of the cross-sectional shape of the roadway on the shape of the plastic zone is related to the size of the load. When the load is small, the influence of the cross-sectional shape is large. When the load increases to a certain extent, the influence of the cross-sectional shape on the shape of the plastic zone gradually weakens.

When the load direction of the surrounding rock of the roadway changes, it will cause the plastic zone of the surrounding rock of the roadway to rotate, resulting in asymmetric damage to the surrounding rock of the roadway, with the depth of roof damage on one side significantly larger than the other. When the ratio of the two-way load of the roadway is large, the change of the load direction will cause the wing angle part of the butterfly-shaped plastic zone to appear above the roof of the roadway, which is easy to produce a falling arch, which is extremely unfavorable for the maintenance of the roadway.

The lithology and combination of the surrounding rock of the roadway also have a significant impact on the geometric size and distribution shape of the plastic zone of the surrounding rock. The range of the plastic zone of the surrounding rock decreases with the increase of the strength of the rock mass. The layered combination rock mass surrounding the roadway can also produce butterfly-shaped and other irregular plastic zones when the ratio of the two-way load is large. A sudden change in the boundary of the plastic zone will occur on the contact surface between the rock layers.

During different stages of mining influence, the formation and development of the plastic zone of the surrounding rock in the mining roadway show different distribution patterns and evolution laws: During the roadway excavation stage, the stress of the surrounding rock near the roadway is redistributed, and the stress of some surrounding rocks increases. When it exceeds the strength limit of the surrounding rock, plastic deformation and damage occur. The range of the plastic zone is small, the boundary of the plastic zone is symmetrical, and the maximum damage depth occurs at the central position of the top and bottom of the roadway and both sides. After being severely affected by mining, the boundaries of the plastic zones of the top, bottom, and both sides of the roadway expand asymmetrically, and the range of plastic zone damage continues to expand. The plastic zone of the surrounding rock in the mining roadway is mainly caused by the mining influence of the coal mining face; due to the influence of the superimposed stress field of mining, the mining stress increases rapidly, and the shape of the plastic zone also changes, forming an irregular "butterfly" plastic zone with a certain rotation angle, as shown in Figs. 5.5 and 5.6.

Generally speaking, the degree of fragmentation of the surrounding rock of the roadway can directly reflect the stability of the surrounding rock. When constructing

Fig. 5.6 Irregular "butterfly" plastic zone

in areas where the surrounding rock is more fragmented, the mechanical properties of the fragmented surrounding rock have the following points:

(1) The overall strength of the fragmented surrounding rock is not high. During the excavation of the roadway, it cannot maintain the stability of the roadway by its own ability, and it is more prone to instability and damage than the intact surrounding rock.
(2) The joints and cracks of the fragmented surrounding rock are more developed than the intact surrounding rock, and their distribution also shows no obvious regularity. Their structural surfaces also show mutual intersections and no significant directivity. Therefore, to a certain extent, it can be regarded as a continuum with the same properties in all directions.
(3) When the fragmented rock mass is approximately regarded as a continuum, the rock mass as a whole will show certain elastoplastic characteristics. This elastoplastic characteristic can be understood as the stress in the rock mass causing the rock mass to fail and produce slippage. After that, it maintains a certain strength with the development of deformation and has strain hardening characteristics. In theoretical analysis, the elastoplastic theory of continuous media should be used as the basis for analysis.

He Manchao and others started with the stress environment where the deep roadway is located, analyzed the mechanical mechanism and characteristics of large deformation in deep mines, and believed that a significant feature of deep mines is high stress level and complex stress mechanism. First, the roadway is buried deep and has a gravity mechanism. The shallow surrounding rock of each rock group enters a deep high-stress state and strain softening state after being exposed.

According to the results of regional tectonic system and in-situ stress field measurement, the azimuth of the main roadway and the angle between the stress direction are large, and the tectonic stress has a significant impact on the deep roadway, so the deformation of the deep roadway has a tectonic stress mechanism.

Deep mining is strongly affected by mining. Due to the characteristics of deep mining and the arrangement of mining sequence, deep roadway engineering is affected by the surrounding working faces during construction and maintenance, and the disturbance frequency of neighboring mining to deep roadways is high and correspondingly strong. The roadway has a mechanism of engineering bias stress.

Fracture development and large rock structure fragmentation are another feature of deep soft rock roadways. From the analysis of diagenesis and coal-bearing construction in the aforementioned mining area, mudstone is soft and brittle, sandy mudstone and sandstone have medium strength, but they are rich in carbonaceous and muddy stripes, bedding joints are developed, and the rock structure has multiple slip surfaces and weak interlayers. From the statistical analysis of core recovery, the core recovery rate and RQD index of mudstone group and sandy mudstone are low, so the overall strength of deep rock mass, especially mudstone group and sandstone group, is low due to the influence of structural surface, and deep soft rock has the characteristics of jointed soft rock.

Mudstone also has strong water softening characteristics and certain water absorption expansion characteristics. In mudstone, the content of clay minerals is high, and water softens the rock body through lubrication, water wedge action, dissolution and potential erosion, causing the strength of the rock body to drop sharply after encountering water. In addition, mudstone has a certain colloidal expansion mechanism.

5.2 Research on Rock Deterioration Model Based on Stress Gradient Theory

The excavation process of underground engineering produces structural free surfaces, causing the stress near the excavation boundary to redistribute, forming a new multi-axial stress state in the surrounding rock. A large amount of research [34–38] It has been confirmed that the confining pressure has a significant impact on the mechanical behavior of rocks, especially in terms of strength and fracture characteristics. The rate of change of confining pressure in a certain direction directly determines the mode of failure of the surrounding rock. Under high ground stress conditions, the radial stress shows a significant gradient change in the surrounding rock due to the influence of vertical stress, which in turn affects the stability of the surrounding rock. After excavation, the rock mass deteriorates and becomes unstable due to different stress environments [39–44]. Instability or damage of the surrounding rock often occurs in areas with significant stress gradients. Therefore, the stress gradient is one of the main factors considered in the analysis of rock mass stability.

5.2.1 Rock Mass Deterioration Parameters Based on Stress Gradient Theory

The existence of natural fractures in the rock mass makes it impossible to satisfy the continuity assumption of classical continuum mechanics [45, 46], considering the rock as an ideal elastoplastic material, assuming that ε_{ij} and σ_{ij} comply with the generalized Hooke's law within the linear elasticity range:

$$\begin{cases} \varepsilon_{i(j,k)} = \frac{1}{E}\left[\sigma_{i(j,k)} - \mu\left(\sigma_{j(k,i)} + \sigma_{k(i,j)}\right)\right] \\ \gamma_{jk(ki,ij)} = \frac{\tau_{jk(ki,ij)}}{G} \end{cases} \quad (5.4)$$

where, $\sigma_{i(j,k)}$ is the stress component; $\varepsilon_{i(j,k)}$ is the strain component; G is the Lame constant; E is the elastic modulus.

The calculation process considers the static process, and the balance equation is as follows:

$$\sigma_{ij,i} - \tau_{ijk,ij} + f_i = 0 \quad (5.5)$$

where, f_i represents body force, that is, the external force received by the physical internal microelement. The stress component is represented by the statistical average of the stress changes on the relatively small but not infinitesimal volume of the rock mass and the stress gradient $T(x, y)$ is calculated,

$$T(x, y) = |\nabla f| = \sqrt{\left(\frac{\partial f}{\partial x}\right)^2 + \left(\frac{\partial f}{\partial y}\right)^2} \quad (5.6)$$

The stress gradient induces discontinuous and gradient-type expansion deformation such as delamination, sliding, and tensile and shear cracks in the surrounding rock of the tunnel [47]. To maintain the integrity and self-supporting ability of the surrounding rock, and to reduce the expansion and degradation of the rock mass caused by stress gradient, the method of reinforcing the surrounding rock with anchor bolts and sprayed concrete is usually used. Therefore, it is proposed to use the stress gradient compensation coefficient to evaluate the effect of supporting the deteriorated rock mass. According to the distribution law of stress gradient, the stress gradient compensation coefficient η can be expressed as:

$$\eta = \frac{(\sigma_{ij}/P_0)dx - (\sigma_0/P_0)dx}{(\sigma_{ij}/P_0)dx}(0 < \eta < 1) \quad (5.7)$$

where, σ_{ij} is the horizontal stress under different pre-tightening support conditions, σ_0 is the horizontal stress of the rock mass under the initial excavation conditions, P_0 is the original rock stress.

5.2 Research on Rock Deterioration Model Based on Stress Gradient Theory

The rock is assumed to be an isotropic composite composed of countless microelements. The spacing and volume of the microelements after anchoring with anchor bolts are smaller than those under the initial excavation state, and the increase in the interaction force between the microelements leads to an increase in the stress gradient (Fig. 5.7). Based on the differences in mechanical parameters in the actual rock mass and the discreteness of rock material properties, the effective stress gradient compensation coefficient $\tilde{\eta}$, and η numerical relationship is as follows:

$$\tilde{\eta} = \frac{T_{(\text{actual value})}}{T_{(\text{theoretical value})}} \cdot \eta \tag{5.8}$$

where, $T_{(\text{actual value})}$ is the actual value of the rock stress gradient, $T_{(\text{theoretical value})}$ is the theoretical value of the rock stress gradient.

Meanwhile, during the process of stress redistribution of the surrounding rock of the roadway, the stability of the surrounding rock significantly decreases. The stress gradient damage variable is used to characterize the stability of the rock mass, and the stability coefficient Y is expressed as:

$$Y = 1 - \frac{(\sigma_{ij}/P_0)dx - T(x)}{(\sigma_{ij}/P_0)dx} \tag{5.9}$$

where, $T(x)$ is the stress gradient.

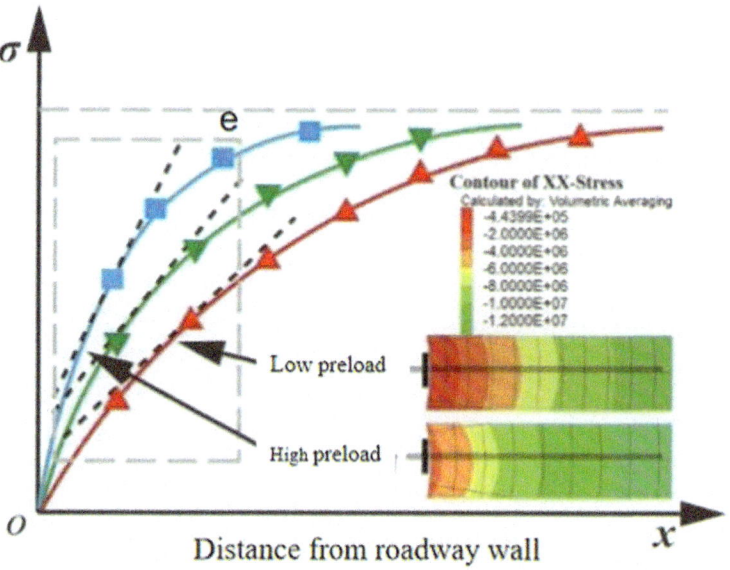

Fig. 5.7 Schematic diagram of stress gradient compensation

5.2.2 Solution of the Rock Mass Deterioration Model Based on Stress Gradient Theory

1. Calculation of the ideal elastic–plastic stress gradient solution based on the M-C criterion

To simplify the research, we assume that the deep-buried circular flat roadway is infinitely long; the original rock stress is isotropic; the rock mass is an ideal elastic–plastic material; the roadway burial depth $\geq 20 R_0$ [48–50]. The constitutive relationship of the plastic zone adopts the incremental constitutive relationship, that is, the Levy–Mises relationship [51] is determined. The mechanical model and constitutive relationship of the axisymmetric circular roadway are shown in Fig. 5.8.

Its elastic solution, according to the research results of predecessors [52–54] have

$$\sigma_r = P_0\left(1 - \frac{R_0^2}{r^2}\right) \tag{5.10}$$

$$\sigma_\theta = P_0\left(1 + \frac{R_0^2}{r^2}\right) \tag{5.11}$$

where, σ_r, σ_θ representing the radial stress and tangential stress of the surrounding rock elastic zone respectively; R_0 represents the radius of the circular tunnel; r represents the radius of any point in the surrounding rock; P_0 represents the original rock stress.

According to the generalized Hooke's law: $\varepsilon_z = \frac{1}{E}[\sigma_z - v(\sigma_r + \sigma_\theta)]$, then in the original rock stress zone after tunnel excavation, there is

$$P_0 = v(P_0 + P_0) + E\varepsilon_z \tag{5.12a}$$

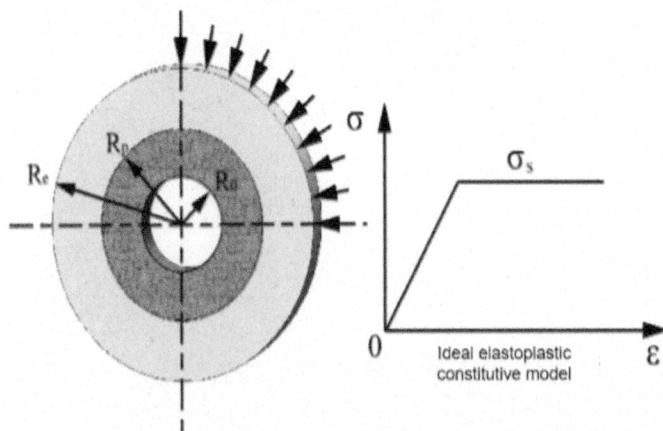

Fig. 5.8 Mechanical model and constitutive relationship of roadway surrounding rock

5.2 Research on Rock Deterioration Model Based on Stress Gradient Theory

where, E represents the elastic modulus, v represents the Poisson's ratio.

In the elastic stress zone after tunnel excavation, for an infinitely long tunnel, it can be simplified as a plane strain problem, there is

$$\sigma_Z = v(\sigma_r + \sigma_\theta) + E\varepsilon_Z \tag{5.12b}$$

Subtracting Eq. 5.12b from 5.12a and rearranging, we get

$$\begin{aligned}\sigma_Z &= v(\sigma_r + \sigma_\theta - 2P_0) + P_0\sigma_Z \\ &= v(\sigma_r + \sigma_\theta) + E\varepsilon_Z\end{aligned} \tag{5.12c}$$

Substituting Eqs. 5.10 and 5.11 into Eq. 5.12c, we get

$$\sigma_Z = P_0 \tag{5.13}$$

where, σ_Z and ε_Z represent the stress and strain in the surrounding rock elastic zone along the tunnel axis direction respectively.

For the original rock stress when the surrounding rock begins to yield, according to the M-C yield condition, there is

$$\tau = c + \sigma_n \tan\varphi \tag{5.14}$$

where, τ and σ_n represent the shear stress and normal stress on the failure surface respectively, c and φ represent the cohesion and friction angle respectively.

Combine it with the Mohr circle to transform it into the following formula:

$$\frac{\sigma_1 - \sigma_3}{2} = c \cdot \cos\varphi + \frac{\sigma_1 + \sigma_3}{2}\sin\varphi \tag{5.15}$$

where, σ_1 and σ_3 represent the maximum compressive stress and minimum compressive stress corresponding to the peak value during the loading process $\frac{\sigma_1-\sigma_3}{2}$ respectively.

Introduce shear stress $\tau_{13} = \frac{\sigma_1-\sigma_3}{2}$, $\eta_{13} = \frac{\sigma_1+\sigma_3}{2}$

$$\sigma_\theta = \frac{2c \cdot \cos\varphi}{1 - \sin\varphi} + \sigma_r \frac{1 + \sin\varphi}{1 - \sin\varphi} = \sigma_C + \sigma_r \cdot \varepsilon \tag{5.16}$$

Its elastic–plastic solution, in the elastic zone, the stress solution of the elastic zone with undetermined integral constant is:

$$\begin{cases}\sigma_r = A - \frac{B}{r^2} \\ \sigma_\theta = A + \frac{B}{r^2}\end{cases} \tag{5.17}$$

In the plastic zone, when the original rock stress $P_0 > P_0^{M-C}$ reaches the M-C yield condition, the yield range of the tunnel surrounding rock, that is, the radius of the elastic–plastic interface, continues to increase. According to the Levy–Mises constitutive relationship, there is:

$$\frac{2\sigma_z - \sigma_r - \sigma_\theta}{2\sigma_r - \sigma_\theta - \sigma_z} = \frac{d\varepsilon_z}{d\varepsilon_r} \tag{5.18}$$

Get:

$$\sigma_z = \frac{\sigma_r + \sigma_\theta}{2} \tag{5.19}$$

Substituting Eq. 5.19 into Eq. 5.14, we get:

$$\sigma_r - \sigma_\theta = (1 - \varepsilon)\sigma_r - \sigma_C \tag{5.20}$$

Substituting Eq. 5.20 into the balance equation, we get

$$\frac{d\sigma_r}{dr} + \frac{\sigma_r - \sigma_\theta}{r} = 0 \tag{5.21}$$

Solve to get:

$$\frac{d\sigma_r}{(\varepsilon - 1)\sigma_r + \sigma_C} = \frac{dr}{r} \tag{5.22}$$

Integrating Eq. 5.22, we get:

$$\sigma_r = C_1 \cdot r^{\varepsilon - 1} - \frac{\sigma_C}{\varepsilon - 1} \tag{5.23}$$

Boundary conditions:
Elastic zone:
Outer boundary

$$r \to \infty; \sigma_r = \sigma_\theta = P_0 \tag{5.24}$$

Inner boundary (elastic–plastic interface):

$$\begin{cases} \left. \begin{array}{c} \sigma_r \\ \sigma_\theta \end{array} \right\} = A \mp \frac{B}{R_p^2} \\ \sigma_z = \frac{\sigma_r + \sigma_\theta}{2} = A \end{cases} \tag{5.25}$$

Plastic zone:
Outer boundary (elastic–plastic interface)

$$\sigma_r^P = \sigma_r^e \tag{5.26}$$

5.2 Research on Rock Deterioration Model Based on Stress Gradient Theory

$$\sigma_\theta^P = \sigma_\theta^e \tag{5.27}$$

Inner boundary $\sigma_r = 0$
So

$$C_1 = \frac{\sigma_C}{\varepsilon - 1} \cdot R_0^{1-\varepsilon} \tag{5.28}$$

Solving Eqs. 5.19, 5.20, 5.23, and 5.28 together, we get the stress in the plastic zone:

$$\sigma_r^P = \frac{\sigma_C}{\varepsilon - 1}\left[\left(\frac{r}{R_0}\right)^{\varepsilon - 1} - 1\right] \tag{5.29}$$

$$\sigma_\theta^P = \frac{\sigma_C \cdot \varepsilon}{\varepsilon - 1}\left[\left(\frac{r}{R_0}\right)^{\varepsilon - 1} - 1\right] + \sigma_C \tag{5.30}$$

$$\sigma_z^P = \frac{\sigma_C \cdot (\varepsilon + 1)}{2(\varepsilon - 1)}\left[\left(\frac{r}{R_0}\right)^{\varepsilon - 1} - 1\right] + \frac{\sigma_C}{2} \tag{5.31}$$

From Eqs. 5.17 and 5.24, we get:

$$\sigma_r = P_0 - \frac{B}{r^2} \tag{5.32}$$

$$\sigma_\theta = P_0 + \frac{B}{r^2} \tag{5.33}$$

Solving Eqs. 5.29, 5.32, and 5.26 together, we get:

$$\frac{\sigma_C}{\varepsilon - 1}\left[\left(\frac{r}{R_0}\right)^{\varepsilon - 1} - 1\right] = P_0 - \frac{B}{P_P^2} \tag{5.34}$$

$$B = P_P^2\left\{P_0 - \frac{\sigma_C}{\varepsilon - 1}\left[\left(\frac{r}{R_0}\right)^{\varepsilon - 1} - 1\right]\right\} \tag{5.35}$$

Solving Eqs. 5.19, 5.32, 5.33, and 5.35 together, we get:

$$\sigma_r^e = P_0 - \left\{P_0 - \frac{\sigma_C}{\varepsilon - 1}\left[\left(\frac{R_P}{R_0}\right)^{\varepsilon - 1} - 1\right]\right\} \tag{5.36}$$

$$\sigma_\theta^e = P_0 + \left\{P_0 - \frac{\sigma_C}{\varepsilon - 1}\left[\left(\frac{R_P}{R_0}\right)^{\varepsilon - 1} - 1\right]\right\} \tag{5.37}$$

$$\sigma_\theta^e = P_0 \tag{5.38}$$

The stress balance relationship at the elastic–plastic interface is obtained from Eqs. 5.27, 5.30, and 5.37:

$$\frac{\sigma_C \cdot \varepsilon}{\varepsilon - 1}\left[\left(\frac{r}{R_0}\right)^{\varepsilon-1} - 1\right] + \sigma_C = 2P_0 - \frac{\sigma_C}{\varepsilon - 1}\left[\left(\frac{R_P}{R_0}\right)^{\varepsilon-1} - 1\right] \tag{5.39}$$

The radius of the plastic zone is obtained by solving Eq. 5.29:

$$R_P = R_0 \cdot \left[\frac{(2P_0 - \sigma_C)(\varepsilon - 1)}{\sigma_C(\varepsilon + 1)} + 1\right]^{\frac{1}{\varepsilon-1}} \tag{5.40}$$

Let $\beta = \left[\frac{(2P_0 - \sigma_C)(\varepsilon-1)}{\sigma_C(\varepsilon+1)} + 1\right]^{\frac{1}{\varepsilon-1}}$, Eq. 5.40 can be simplified to $R_P = R_0 \cdot \beta$.

The stress in the elastic zone is obtained by solving Eqs. 5.32, 5.33, 5.35, and 5.40 together:

$$\begin{cases} \sigma_r^e \\ \sigma_\theta^e \end{cases} = P_0 \mp \frac{R_0^2}{r^2}\beta^2\left[P_0 - \frac{\sigma_C}{\varepsilon-1}\left(\beta^{\varepsilon-1} - 1\right)\right] \\ \sigma_Z^e = P_0 \tag{5.41}$$

$$\begin{cases} \sigma_r^e \\ \sigma_\theta^e \end{cases} = P_0 \mp N\frac{1}{r^2} \\ \sigma_Z^e = P_0 \tag{5.42}$$

Let $N = R_0^2 \beta^2\left[P_0 - \frac{\sigma_C}{\varepsilon-1}\left(\beta^{\varepsilon-1} - 1\right)\right]$, then Eq. 5.41 can be simplified to Convert polar coordinates to Cartesian coordinates:

$$\sigma_r^e = P_0 - N\frac{1}{x^2 + y^2}$$

$$\sigma_\theta^e = P_0 + N\frac{1}{x^2 + y^2}$$

$$\sigma_x = \sigma_r \cdot \frac{x^2}{x^2 + y^2} + \sigma_\theta \cdot \frac{y^2}{x^2 + y^2} = P_0 - N\frac{x^2 - y^2}{(x^2 + y^2)^2}$$

$$\sigma_y = \sigma_r \cdot \frac{y^2}{x^2 + y^2} + \sigma_\theta \cdot \frac{x^2}{x^2 + y^2} = P_0 + N\frac{x^2 - y^2}{(x^2 + y^2)^2}$$

$$\tau_{xy} = -2N\frac{xy}{(x^2 + y^2)^2} \tag{5.43}$$

Have

5.2 Research on Rock Deterioration Model Based on Stress Gradient Theory

$$\begin{cases} \frac{\sigma_1 - \sigma_3}{2} = c \cdot \cos\varphi + \frac{\sigma_1 + \sigma_3}{2} \sin\varphi \\ \begin{Bmatrix} \sigma_1 \\ \sigma_3 \end{Bmatrix} = \frac{\sigma_x + \sigma_y}{2} \pm \sqrt{\left(\frac{\sigma_x - \sigma_y}{2}\right)^2 + \tau_{xy}^2} \end{cases} \quad (5.44)$$

According to Eq. 5.14, set

$$f(x, y, z) = \tau - c + \sigma_n \tan\varphi \quad (5.45)$$

Another expression based on the M-C criterion can be obtained:

$$f(x, y, z) = \tau - c + \sigma_n \tan\varphi = 0$$

Substitute Eq. 5.44 into Eq. 5.45 to get

$$f(x, y, z) = 2\sqrt{\left(\frac{\sigma_x - \sigma_y}{2}\right)^2 + \tau_{xy}^2} - \frac{\sigma_x + \sigma_y}{2} \cdot \sin\varphi - c \cdot \cos\varphi = 0 \quad (5.46)$$

Calculate the gradient field of formula 5.46 to get:

$$\nabla f = \left(\frac{\partial f}{\partial x}\vec{i} + \frac{\partial f}{\partial y}\vec{j} + \frac{\partial f}{\partial z}\vec{k}\right) \quad (5.47)$$

Formula 5.47 is the stress field gradient based on the M-C criterion. The stress field of the surrounding rock in the tunnel is considered as a plane strain problem, so formula 5.47 can be simplified to:

$$\nabla f = \left(\frac{\partial f}{\partial x}\vec{i} + \frac{\partial f}{\partial y}\vec{j}\right) \quad (5.48)$$

Under plane strain conditions, the stress field gradient value based on the M-C criterion is:

$$\nabla f = \left(\frac{\partial f}{\partial x}\vec{i} + \frac{\partial f}{\partial y}\vec{j} + \frac{\partial f}{\partial z}\vec{k}\right) T(x, y) = |\nabla f|$$
$$= \sqrt{\left(\frac{\partial f}{\partial x}\right)^2 + \left(\frac{\partial f}{\partial y}\right)^2} \quad (5.49)$$

Solution of stress gradient in elastic zone:

$$\frac{\partial \sigma_x}{\partial x} = N \cdot \frac{2x(x^2 - 3y^2)}{(x^2 + y^2)^3}; \quad \frac{\partial \sigma_y}{\partial x} = -N \cdot \frac{2x(x^2 - 3y^2)}{(x^2 + y^2)^3};$$

$$\frac{\partial \sigma_x}{\partial y} = N \cdot \frac{2y(3x^2 - y^2)}{(x^2 + y^2)^3}; \quad \frac{\partial \sigma_y}{\partial y} = -N \cdot \frac{2y(3x^2 - y^2)}{(x^2 + y^2)^3};$$

$$\frac{\partial \tau_{xy}}{\partial x} = -2N \cdot \frac{y(y^2 - 3x^2)}{(x^2 + y^2)^3}; \quad \frac{\partial \tau_{xy}}{\partial y} = -2N \cdot \frac{x(x^2 - 3y^2)}{(x^2 + y^2)^3};$$

Solution of stress gradient in plastic zone:

$$\frac{\partial \sigma_x}{\partial x} = \frac{\partial \sigma_r}{\partial x} \cdot \frac{x^2}{x^2 + y^2} + \sigma_r \cdot \frac{2x(x^2 + y^2) - 2x^3}{(x^2 + y^2)^3} - \frac{2xy^2 \cdot \sigma_\theta}{(x^2 + y^2)^2} + \frac{\partial \sigma_\theta}{\partial x} \cdot \frac{y^2}{x^2 + y^2};$$

$$\frac{\partial \sigma_y}{\partial x} = \frac{\partial \sigma_r}{\partial x} \cdot \frac{y^2}{x^2 + y^2} - \sigma_r \cdot \frac{2xy^2}{(x^2 + y^2)^2} + \sigma_\theta \cdot \frac{2x(x^2 + y^2) - 2x^3}{(x^2 + y^2)^2} + \frac{\partial \sigma_\theta}{\partial x} \cdot \frac{x^2}{x^2 + y^2};$$

$$\frac{\partial \sigma_x}{\partial y} = \sigma_r \cdot \frac{-x^2 2y}{(x^2 + y^2)^2} + \frac{\partial \sigma_r}{\partial y} \cdot \frac{x^2}{x^2 + y^2} + \sigma_\theta \cdot \frac{2y(x^2 + y^2) - 2y^3}{(x^2 + y^2)^2} + \frac{\partial \sigma_\theta}{\partial y} \cdot \frac{y^2}{x^2 + y^2};$$

$$\frac{\partial \sigma_y}{\partial y} = \sigma_r \cdot \frac{2y(x^2 + y^2) - 2y^3}{(x^2 + y^2)^2} + \frac{\partial \sigma_r}{\partial y} \cdot \frac{y^2}{x^2 + y^2} + \sigma_\theta \cdot \frac{-x^2 2y}{(x^2 + y^2)^2} + \frac{\partial \sigma_\theta}{\partial y} \cdot \frac{x^2}{x^2 + y^2};$$

$$\frac{\partial \tau_{xy}}{\partial x} = \left(\frac{\partial \sigma_r}{\partial x} - \frac{\partial \sigma_\theta}{\partial x}\right) \frac{xy}{x^2 + y^2} + (\sigma_r - \sigma_\theta) \cdot \frac{y(x^2 + y^2) - xy2x}{(x^2 + y^2)^2};$$

$$\frac{\partial \tau_{xy}}{\partial y} = \left(\frac{\partial \sigma_r}{\partial y} - \frac{\partial \sigma_\theta}{\partial y}\right) \frac{xy}{x^2 + y^2} + (\sigma_r - \sigma_\theta) \cdot \frac{x(x^2 + y^2) - xy2y}{(x^2 + y^2)^2};$$

$$\frac{\partial \sigma_r^p}{\partial x} = \frac{x \cdot c \cdot \cos\varphi(1 - \sin\varphi)}{2\sin\varphi^2} \left(\frac{x^2 + y^2}{R_0^2}\right)^{\frac{1 - 5\sin\varphi}{4\sin\varphi}};$$

$$\frac{\partial \sigma_\theta^p}{\partial x} = \frac{x \cdot c \cdot \cos\varphi(1 + \sin\varphi)}{2\sin\varphi^2} \left(\frac{x^2 + y^2}{R_0^2}\right)^{\frac{1 - 5\sin\varphi}{4\sin\varphi}};$$

$$\frac{\partial \sigma_r^p}{\partial y} = \frac{y \cdot c \cdot \cos\varphi(1 - \sin\varphi)}{2\sin\varphi^2} \left(\frac{x^2 + y^2}{R_0^2}\right)^{\frac{1 - 5\sin\varphi}{4\sin\varphi}};$$

$$\frac{\partial \sigma_\theta^p}{\partial y} = \frac{y \cdot c \cdot \cos\varphi(1 + \sin\varphi)}{2\sin\varphi^2} \left(\frac{x^2 + y^2}{R_0^2}\right)^{\frac{1 - 5\sin\varphi}{4\sin\varphi}};$$

2. Calculation of ideal elastic–plastic stress gradient solution based on D-P criterion

The calculation process is consistent with the M-C criterion, and the solution results are listed below.

According to the D-P yield condition, we have

$$\sqrt{J_2} - \alpha I_1 - k = 0 \tag{5.50}$$

5.2 Research on Rock Deterioration Model Based on Stress Gradient Theory

The stress in the plastic zone is calculated as:

$$\sigma_r^P = \frac{k}{3\alpha}\left[\left(\frac{r}{R_0}\right)^{\frac{6\alpha}{1-3\alpha}} - 1\right] \tag{5.51}$$

$$\sigma_\theta^P = \frac{k}{3\alpha}\left[\frac{1+3\alpha}{1-3\alpha}\left(\frac{r}{R_0}\right)^{\frac{6\alpha}{1-3\alpha}} - 1\right] \tag{5.52}$$

$$\sigma_Z^P = \frac{k}{3\alpha}\left[\frac{1}{1-3\alpha}\left(\frac{r}{R_0}\right)^{\frac{6\alpha}{1-3\alpha}} - 1\right] \tag{5.53}$$

The stress in the elastic zone is calculated as:

$$\begin{cases} \sigma_r^e = P_0 - \frac{R_0^2}{r^2}k(1-3\alpha)^{\frac{1-3\alpha}{3\alpha}}\left(1 + \frac{3\alpha}{k}P_0\right)^{\frac{1}{3\alpha}} \\ \sigma_\theta^e = P_0 + -\frac{R_0^2}{r^2}k(1-3\alpha)^{\frac{1-3\alpha}{3\alpha}}\left(1 + \frac{3\alpha}{k}P_0\right)^{\frac{1}{3\alpha}} \\ \sigma_Z^P = P_0 \end{cases} \tag{5.54}$$

The stress field gradient value based on the D-P criterion is:

$$T(x,y) = |\nabla f|$$
$$= \sqrt{\left[\frac{1}{2}\left(\sqrt{J_2'}\right)^{-\frac{1}{2}}\frac{\partial J_2'}{\partial x} - \alpha\frac{\partial I_1}{\partial x}\right]^2 + \left[\frac{1}{2}\left(\sqrt{J_2'}\right)^{-\frac{1}{2}}\frac{\partial J_2'}{\partial y} - \alpha\frac{\partial I_1}{\partial y}\right]^2} \tag{5.55}$$

3. Calculation of ideal elastic–plastic stress gradient solution based on H-B criterion

The calculation process is consistent with the M-C criterion, and the solution results are listed below.

According to the H-B yield condition, we have

$$\sigma_1 = \sigma_3 + \sqrt{m\sigma_C\sigma_3 + s \cdot \sigma_C^2} \tag{5.56}$$

The calculated stress in the plastic zone is

$$\begin{cases} \sigma_r^P = \sqrt{s}In\frac{r}{R_0} + \frac{m\sigma_C}{4}\left(In\frac{r}{R_0}\right)^2 \\ \sigma_\theta^P = \sqrt{s}\sigma_C + \left(\frac{m\sigma_C}{2} + \sqrt{s}\sigma_C\right)In\frac{r}{R_0} + \frac{m\sigma_C}{4}\left(In\frac{r}{R_0}\right)^2 \\ \sigma_Z^P = \frac{\sqrt{s}\sigma_C}{2} + \left(\frac{m\sigma_C}{4} + \sqrt{s}\sigma_C\right)In\frac{r}{R_0} + \frac{m\sigma_C}{4}\left(In\frac{r}{R_0}\right)^2 \end{cases} \tag{5.57}$$

The stress in the elastic zone is

$$\begin{cases} \sigma_r^P = P_0 - \frac{R_0^2}{r^2} N e^{2M} \\ \sigma_\theta^P = P_0 + \frac{R_0^2}{r^2} N e^{2M} \\ \sigma_z^P = P_0 \end{cases} \quad (5.58)$$

where, $M = \frac{\sqrt{m^2 \sigma_C^2 + 16 s \sigma_C^2 + 16 m \sigma_C P_0} - m \sigma_C - 4\sqrt{s}\sigma_C}{2m\sigma_C} N = P_0 - \sqrt{s}\sigma_C M - m\sigma_C \frac{M^2}{4}$.

Under plane strain conditions, the stress field gradient is:

$$T(x, y) = |\nabla f| = \sqrt{\left(\frac{\partial f}{\partial x}\right)^2 + \left(\frac{\partial f}{\partial y}\right)^2} \quad (5.59)$$

The stress gradient change is a phenomenon that is commonly present in the surrounding rock of the roadway. Theoretical studies on the stress gradient of the surrounding rock of the roadway should start from the different stress balance relationships in the plastic and elastic zones, and select the appropriate rock strength theory according to different surrounding rock conditions. Based on the stress gradient theory and the continuous damage mechanics model, the mechanical balance equation and boundary conditions at the interface between the plastic and elastic zones under plane strain conditions are derived, and the theoretical solution methods for the stress gradient of the surrounding rock of the roadway under the Mohr–Coulomb criterion, Drucker-Prager criterion, and Hoek–Brown criterion are proposed.

5.3 Determination of Reasonable Pre-tightening Force of Anchor Bolt Based on Stress Gradient

Based on the research of rock deterioration parameters and rock deterioration model based on stress gradient theory in Sect. 5.2, relying on the field engineering practice of Yangchangwan Coal Mine, taking the Mohr–Coulomb criterion as an example, the simulation results and theoretical calculation values are compared with the help of FLAC3D numerical software. The consistency is 93% and the trend is consistent, which verifies the applicability of the stress gradient solution method proposed in this paper. The research results provide a research idea for determining the reasonable support strength of anchor bolts.

5.3.1 Determination of Reasonable Pre-tightening Force of Anchor Bolt by Theoretical Calculation

The Yangchangwan Coal Mine 130,205 track chute is buried at a depth of −495 m. The deformation of the two sides of the chute is severe, which affects the normal production of the mine. It is necessary to reasonably analyze the anchoring support strength of the surrounding rock to ensure the normal use of the roadway.

Assign the actual rock mechanical parameters to the constructed stress gradient solution model. Taking the Mohr–Coulomb criterion as an example, its calculation parameters are as follows: tunnel radius $R_0 = 2.5$ m, horizontal original rock stress $P_0 = 12.8$ MPa, surrounding rock elastic modulus $E = 5$ GPa, shear modulus $G = 2.14$ GPa, Poisson's ratio $\mu = 0.17$, cohesion $C = 6$ MPa. Using the above calculation parameters, the radial stress gradient of the surrounding rock of the roadway under unsupported conditions based on the Mohr–Coulomb criterion is obtained according to the solution process in Sect. 5.2 (Fig. 5.13), where the radius of the plastic zone is $r_P = 3.25$ m; that is, the surrounding rock is in a post-peak plastic state in the range of 2.5 m $\leq$ r $\leq$ 3.25 m, and the surrounding rock is in a pre-peak elastic state in the range of 3.25 m $\leq$ r $\leq$ 10 m.

5.3.2 Determination of Reasonable Pre-tightening Force of Anchor Bolt by Numerical Simulation

A test section near the heading of the 130,205 working face track chute is selected to use anchor bolts to reinforce the surrounding rock, with a spacing of 600 × 600 mm, and the pre-tightening force of each 3 rows of anchor bolts increases by 25 kN, and the pre-tightening force threshold is 0–200 kN. After the anchored rock body is stabilized, monitor the horizontal stress of the surrounding rock under different pre-tightening forces. The data record is shown in Fig. 5.9, and the calculated actual stress gradient is shown in Fig. 5.10.

The analysis of monitoring data shows that the horizontal stress gradient gradually decreases along the radial direction of the surrounding rock and eventually tends to 0. Taking the analysis of the monitoring data of the 25 kN support pre-tightening force as an example, the same monitoring path will show a gradient change inflection point from shallow to deep. Before this inflection point, the stress gradient gradually increases with the increase of pre-tightening force, and after the inflection point, the stress gradient gradually decreases with the increase of pre-tightening force.

The size of the numerical model is 50 m in the x direction; 0.5 m in the y direction; 50 m in the z direction, the model grid is divided into 33,360 elements; the middle of the model is divided into dense areas, the dense area is located in the center of the model, the radius is 10 m, and the dense area grid is divided into 24,000 elements, the mechanical parameters are shown in Table 5.1, the numerical model is shown in Fig. 5.11, and the simulation results of the stress gradient of the surrounding rock under different anchor bolt pre-tightening forces are shown in Fig. 5.12. It is known from the analysis that the stress gradient of the surrounding rock of the roadway under unsupported conditions gradually increases from the roadway wall to the interior of the rock body to the boundary of the plastic zone, and after crossing the boundary, the stress gradient gradually decreases to 0.

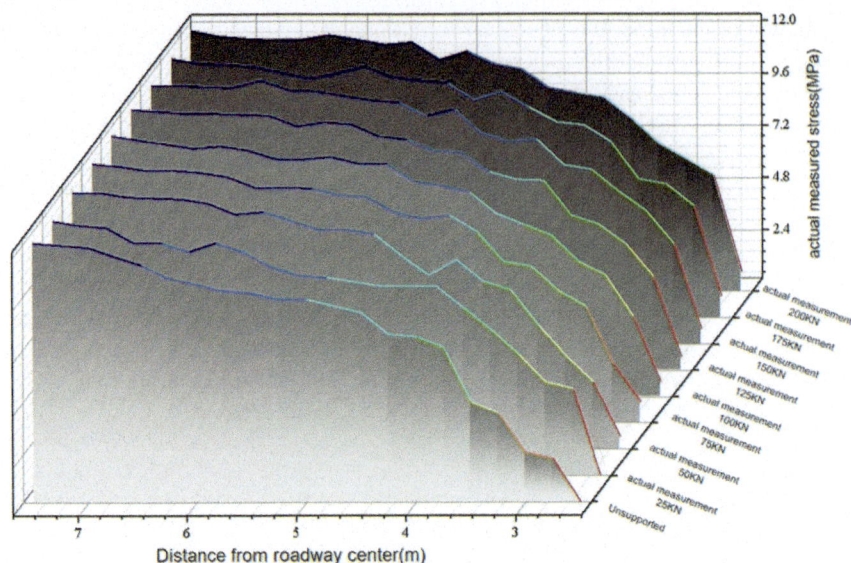

Fig. 5.9 Measured horizontal stress

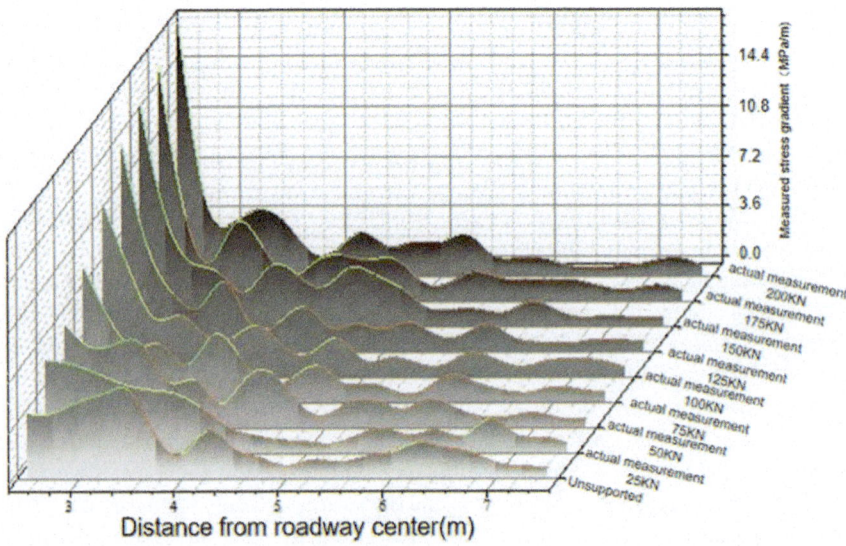

Fig. 5.10 Measured horizontal stress gradient

5.3 Determination of Reasonable Pre-tightening Force of Anchor Bolt …

Table 5.1 Material parameters

Mechanical parameters	Numerical value
Elastic modulus (E, GPa)	5
Poisson's ratio (μ)	0.17
Shear modulus (G, GPa)	2.14
Gravitational density (γ, kN/m^3)	26
Internal friction angle (φ, °)	25
Cohesion (C, MPa)	6
Tensile strength (σ_t, MPa)	7

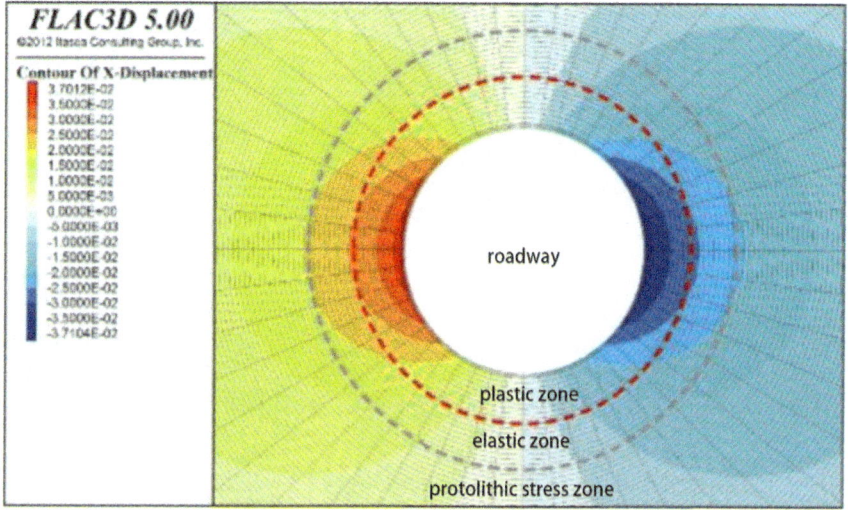

Fig. 5.11 Model schematic

As the pre-tension force increases from 25 to 200 kN, the peak of the horizontal stress gradient in the plastic zone gradually increases from 8 to 15.5 MPa/m, and the range of the plastic zone reduces from 3.5 to 3 m. It can be seen that the increase in the anchor pre-tension force reduces the range of surrounding rock deterioration and shifts the boundary of the plastic zone to the outside of the surrounding rock. The inflection point of the stress gradient change almost coincides with the interface of the surrounding rock elastic–plastic, and will gradually shift to the shallow part of the roadway surrounding rock with the increase of the anchor pre-tension force, and the range of the plastic zone will also shrink, significantly improving the stability of the roadway surrounding rock. According to the proposed surrounding rock stability coefficient Y solving conditions, it is concluded that in the shallow part of the roadway surrounding rock (r < 10 m), the surrounding rock stability coefficient Y is positively correlated with the stress gradient (x), which verifies the effectiveness of the proposed surrounding rock stability model evaluation method.

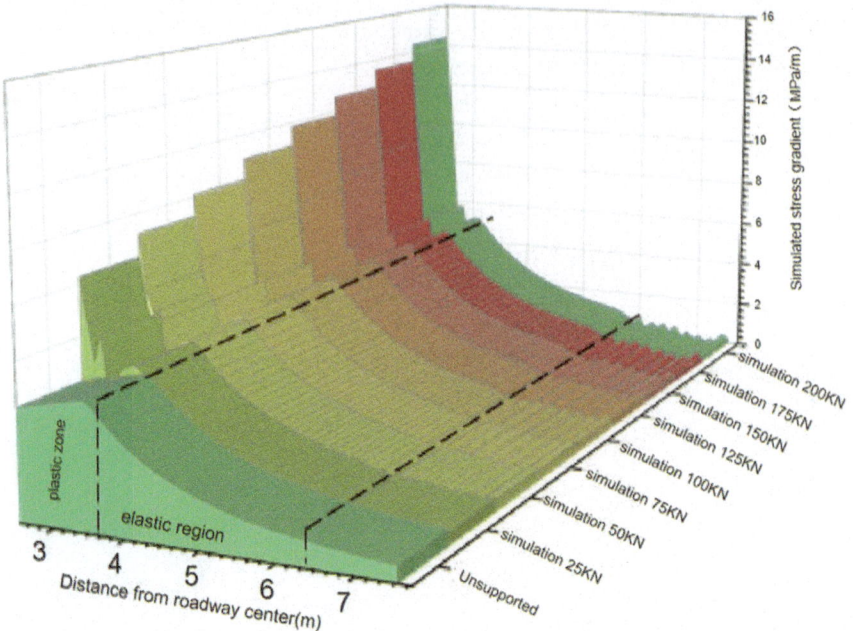

Fig. 5.12 Simulated horizontal stress gradient

5.3.3 Analysis of Theoretical and Measured Correlation

A comparative analysis of the numerical simulation of tunnel deformation under unsupported conditions, theoretical calculations, and field measurement results (Fig. 5.13) is conducted. The correlation analysis of simulated values, theoretical values, and measured values is shown in Table 5.2.

From Table 5.2, it can be concluded that the correlation of the simulated value is 0.93013 > 0.9, and the correlation of the theoretical value is 0.96327 > 0.9. Therefore, the stress gradient of the surrounding rock of the tunnel can be solved by the second-order fitting method. The relationship is as follows:

$$\begin{cases} T_{v(f)} = T_v + B_{1v}x + B_{2v}x^2 \\ T_{t(f)} = T_t + B_{1t}x + B_{2t}x^2 \\ T_{a(f)} = T_a + B_{1a}x + B_{2a}x^2 \end{cases} \quad (5.60)$$

The curve is corrected to

$$T_{t(f)} = T_v + B_{1v}x + 0.9B_{2v}x^2 = T_a + B_{1a}x + 1.2B_{2a}x^2 \quad (5.61)$$

where, v is 'value of simulation'; f is 'fitting values'; t is 'theoretical value'; a is 'actual value'.

5.3 Determination of Reasonable Pre-tightening Force of Anchor Bolt ...

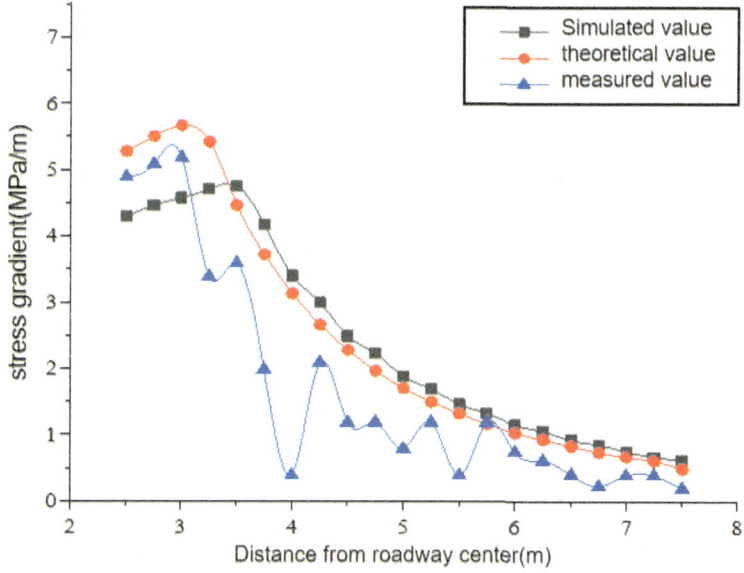

Fig. 5.13 Comparison of theoretical, simulated, and field monitoring results

Table 5.2 Correlation analysis of simulated values, theoretical values, and measured values

		Simulated value	Theoretical value	Measured value
Intercept	Value	9.39326	13.43526	13.62099
	Standard error	1.0362	0.89596	1.58241
B1	Value	−1.92999	−3.49137	−4.13314
	Standard error	0.43974	0.38023	0.67155
B2	Value	0.09759	0.23694	0.32045
	Standard error	0.04358	0.03768	0.06655
Statistics	R^2	0.93013	0.96327	0.86497

Formula 5.61 can be used for the solution between theoretical values, simulated values, and measured values, and for the correct evaluation of measured values based on theoretical values. After obtaining the stress gradient of the surrounding rock under unsupported conditions, with the help of the simulation calculation results (Fig. 5.14b), it is known that when the anchor bolt pretension is 150 kN, the deformation of the surrounding rock of the tunnel under this support form is reduced by 40% compared to the unsupported condition, that is, the tunnel deformation compensation ratio is 40%; the plastic zone range is reduced by 60.4%. The calculation method of the stress gradient compensation coefficient is as follows.

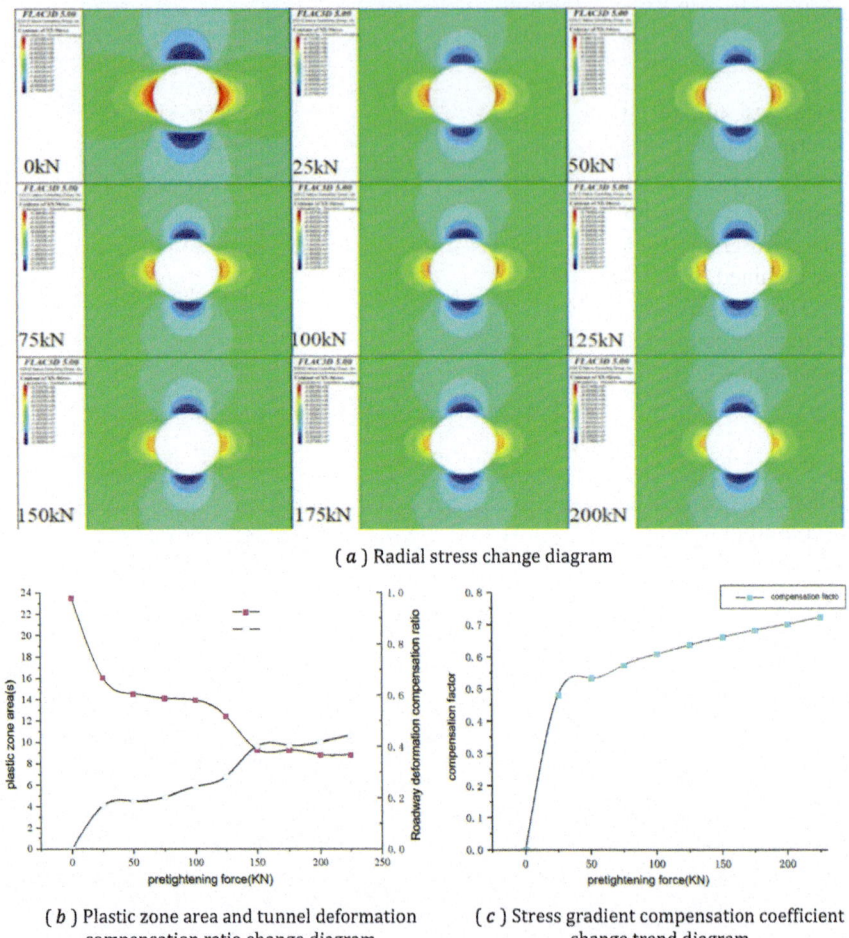

(a) Radial stress change diagram

(b) Plastic zone area and tunnel deformation compensation ratio change diagram

(c) Stress gradient compensation coefficient change trend diagram

Fig. 5.14 Stress gradient compensation effect

The stress gradient benchmark obtained by the theoretical method is T_0, and the stress gradient of the surrounding rock under different pretension anchor bolt support states is T_X, then the compensation coefficient calculation criterion is:

$$\eta = \frac{T_X - T_0}{T_X} \tag{5.62}$$

Effective compensation coefficient:

5.3 Determination of Reasonable Pre-tightening Force of Anchor Bolt ...

$$\tilde{\eta} = \frac{T_{a(f)}}{T_{t(f)}}\eta = 0.65 \cdot \frac{T_a + B_{1a}x + B_{2a}x^2}{T_t + B_{1t}x + B_{2t}x^2} = 0.604 \quad (5.63)$$

where, f is 'fitting values'; t is 'theoretical value'; a is 'actual value'.

The field stress gradient discount value is $(0.65–0.604)/0.65 = 0.07$. Based on the numerical simulation stress gradient compensation coefficient, the corresponding stress gradient compensation effect (Fig. 5.14a) and the change trend of the stress gradient compensation coefficient under different pretensions (Fig. 5.14c) are obtained.

The compensation of the surrounding rock stress gradient by the anchor bolt pretension is positively correlated (Fig. 5.14c). The compensation coefficient obtained by linear fitting of the curve segment is as follows:

$$\eta = 0.00116F_n + 0.47717 \left(R^2 = 0.96427 \right) \quad (5.64)$$

where, F_n is the pretension.

Based on this calculation criterion, the most suitable pretension for anchor bolt support under field conditions can be accurately calculated. The calculation and analysis process is shown in Fig. 5.15.

According to the theoretical analysis results, the stress gradient compensation coefficient 0.65 is selected as the optimal compensation ratio, corresponding to a pretension of 150 kN. According to the estimated discount ratio of 0.07 of the field measured value compared to the theoretical calculation value under this geological condition, the optimal support strength of the anchor bolt design needs to apply a pretension of 161 kN. In the 130,205 track chute, three sets of tunnel surrounding rock deformation and stress monitoring areas are set up. The monitoring equipment is shown in Fig. 5.16. The support optimization effect after construction is shown in Fig. 5.17.

Based on the proposed stress gradient compensation method, under the same surrounding rock conditions, the control effect of tunnel deformation is improved by more than 40%, and the stress gradient compensation coefficient is higher than 0.65, which is higher than the safety control requirements of this mine's surrounding rock deformation. In summary, based on the Mohr–Coulomb strength theory and the tunnel surrounding rock radial stress gradient theory model solution proposed under the guidance of this theory, the numerical simulation and field measured data are basically consistent, and the change trend is consistent, which provides a calculation basis for determining the stress gradient and change law under field conditions; at the same time, according to the change law of the tunnel surrounding rock stress gradient under different pretensions, the stress gradient compensation value and anchor bolt pretension can be optimized to improve the support parameters.

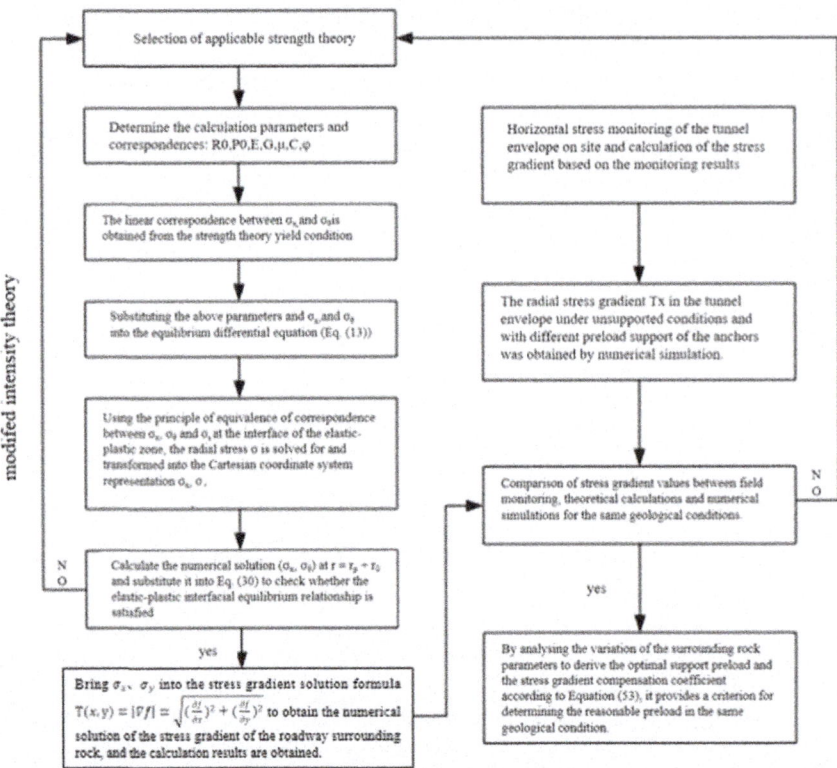

Fig. 5.15 Stress gradient calculation and analysis process

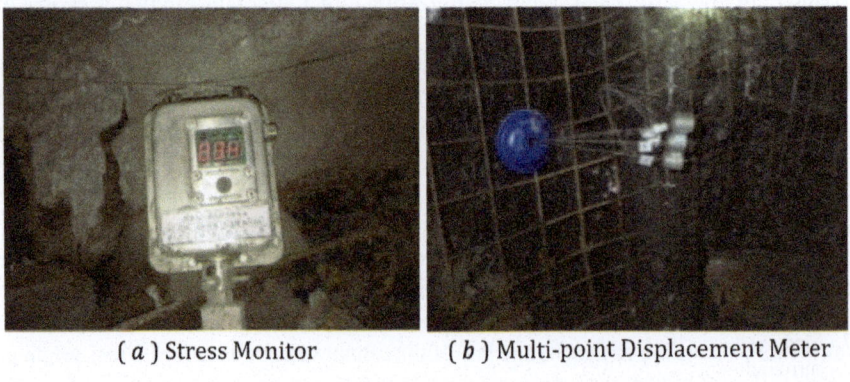

Fig. 5.16 Surrounding rock monitoring equipment

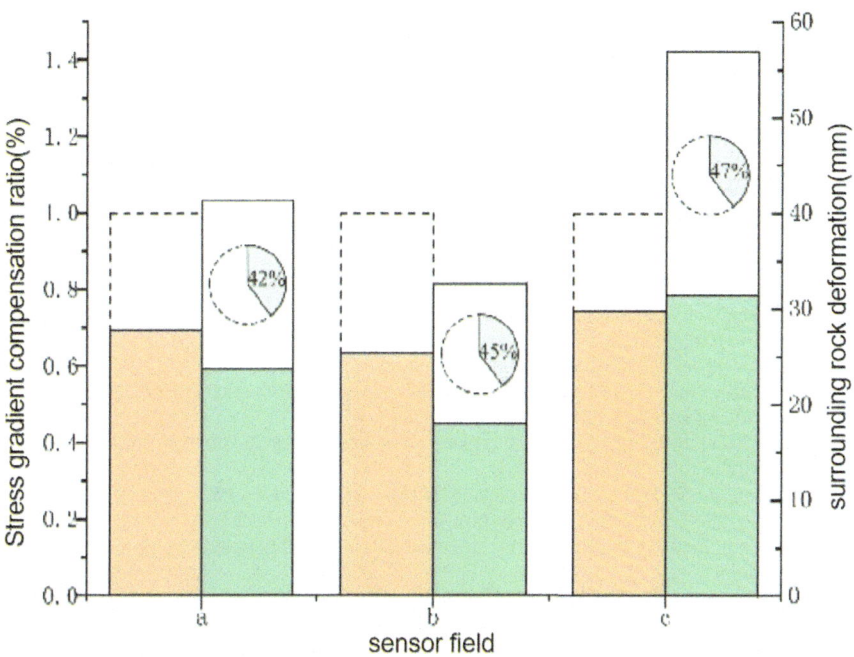

Fig. 5.17 Optimization effect statistics chart

References

1. Xie P, Wu Y, Wang H et al (2013) Mechanism of deformation and destruction of soft rock roadway along empty steeply inclined coal seam group under repeated mining. J Liaoning Tech Univ (Nat Sci Ed) 32(01):44–49
2. Bae GJ, Chang SH, Lee SW et al (2004) Evaluation of interfacial properties between rock mass and shotcrete. Int J Rock Mech Min Sci 41(3):106–112
3. Zhong Q, Yan Z, Guan X (2015) Grouting reinforcement technology of roadway surrounding rock. Coal Technol 34(01):80–83
4. Cao S, Liu C (2004) Grouting reinforcement technology of roof in fault fracture zone of high-grade working face. J Coal (05):545–549
5. Wang X (2017) Research on stability analysis and control technology of bottom structure in broken ore body mining field. Xi'an Univ Arch Technol
6. Wang H (2008) Research on grouting reinforcement and support technology of loose and broken surrounding rock in recovery roadway. China University of Mining and Technology
7. Meng Q, Qian W, Han L et al (2019) Stability control technology and application of roadway surrounding rock in weak ore body. J Min Saf Eng 36(05):906–915
8. Tan Y (2009) Research on determination and support strategy of loose circle in recovery roadway in complex structure belt. Anhui University of Science and Technology
9. Bai J, Hou Z (2006) Research on control principle and application of deep roadway surrounding rock. J China Univ Min Technol (02):145–148
10. Zhao H, Xiong Z, Wang W (2010) Main problems and countermeasures faced by deep mining face in mine. Coal Eng (07):11–13
11. Yuan L, Xue J, Liu Q et al (2011) Theory and support technology of deep rock roadway surrounding rock control in coal mine. J Coal 36(04):535–543

12. Gao Y, Fan Q, Cui X et al (2000) Experimental study on rock rheology and its disturbance effect. Science Press, Beijing
13. Zhang G, Li F (2009) Mine surrounding rock control and disaster prevention. China University of Mining and Technology Press
14. Zhang X (2007) Study on the deformation mechanism and pressure relief control of roadway surrounding rock under dynamic pressure. Anhui University of Science and Technology
15. Jin H (2011) Study on the deformation characteristics of the stope floor and the control of the surrounding rock of the floor roadway. Anhui University of Science and Technology
16. Chen W, Tan X, Lv S et al (2009) Large-scale triaxial compression rheological test and constitutive model study of deep soft rock. Chin J Rock Mech Eng 28(09):1735–1744
17. Lv Y (2016) Modification experiment and application research of grouting support material for broken surrounding rock. Southwest Univ Sci Technol
18. Li J (2012) Study on the anchor grouting support technology for broken surrounding rock of backfill roadway in Dongrong II Mine. Liaoning Tech Univ
19. Su X (2014) Study on the stability characteristics and control technology of broken surrounding rock in deep roadway. Taiyuan Univ Technol
20. Hu F (2012) Stability analysis of broken surrounding rock in Mayazi Tunnel. Chang'an University
21. Li J, Lian M (2011) Mine rock mechanics. Metallurgical Industry Press, Beijing
22. Chen X (2014) Study on the failure mechanism and support technology of the surrounding rock of the inclined shaft under the influence of mining. Hunan University of Science and Technology
23. Liu Q, Zhang H, Lin T (2004) Stability and support countermeasures of deep rock roadway surrounding rock in coal mine. Chin J Rock Mech Eng (21):3732–3737
24. Li Q (2010) Study on the instability mechanism and support technology of roadway under dynamic pressure. Anhui University of Science and Technology
25. Gao Z, Meng X, Fu Z (2014) Elastic-plastic analysis of roadway surrounding rock considering seepage, strain softening and dilation. J Chongqing Univ 37(01):96–101
26. Song J (2002) Empirical strength criterion of rock mass and its application in geological engineering. Geological Publishing House
27. Li XB, Lok TS, Zhao J (2005) Dynamic characteristics of granite subjected to intermediate loading rate. Rock Mech Rock Eng 38(1):21–39
28. Ministry of Construction of the People's Republic of China (1995) GB50218-94 engineering rock mass classification standard. China Architecture & Building Press, Beijing
29. Kang H (2013) Stress distribution characteristics and roadway surrounding rock control technology in deep coal mines. Coal Sci Technol 41(9):12–17
30. Fang X, He J, He J (2009) Research on reinforcement technology of deep high-stress soft rock dynamic pressure roadway. Rock Soil Mech 30(06):1693–1698
31. Fang X, Zhao J, Hong M (2012) Research on deformation mechanism and control of deep well broken surrounding rock roadway. J Min Saf Eng 29(01):1–7
32. Zhao X (2017) Study on the mechanical mechanism of the start of butterfly coal and gas outburst in the excavation roadway. China University of Mining and Technology (Beijing)
33. Zhao Z, Ma N, Liu H et al (2018) Theory of roadway butterfly damage and its application prospects. J China Univ Min Technol 47(05):969–978
34. Alam AKMB, Niioka M, Fujii Y et al (2014) Effects of confining pressure on the permeability of three rock types under compression. Int J Rock Mech Min Sci 65(1):49–61
35. Sukplum W, Wannakao L (2016) Influence of confining pressure on the mechanical behavior of Phu Kradung sandstone. Int J Rock Mech Min Sci 86:48–54
36. Li X B et al (2017) Spalling strength of rock under different static pre-confining pressures. Int J Impact Eng 99(2017):69–74
37. Schmidt RA, Huddle CW (1977) Effect of confining pressure on fracture toughness of Indiana limestone. Int J Rock Mech Min Sci Geomech Abstracts 14(5):289–293
38. Chi TN, Nguyen GD, Das A et al (2017) Constitutive modelling of progressive localised failure in porous sandstones under shearing at high confining pressures. Int J Rock Mech Min Sci 93:179–195

References

39. Barton N, Lien R, Lunde J (1974) Engineering classification of rock masses for the design of tunnel support. Rock Mech 6(4):189–236
40. Cai Y, Esaki T, Jiang Y (2004) A rock bolt and rock mass interaction model. Int J Rock Mech Min Sci 41(7):1055–1067
41. Moore ID (1994) Analysis of rib supports for circular tunnels in elastic ground. Rock Mech Rock Eng 27(3):155–172
42. Kang HP (2014) Support technologies for deep and complex roadways in underground coal mines: a review. Int J Coal Sci Technol 1(3):261–277
43. Bobet A (2009) Elastic solution for deep tunnels application to excavation damage zone and rockbolt support. Rock Mech Rock Eng 42(2):147–174
44. Liu GF, Feng XT, Feng GL et al (2016) A method for dynamic risk assessment and management of rock bursts in drill and blast tunnels. Rock Mech Rock Eng 49(8):3257–3279
45. Barpi F, Valente S, Cravero M et al (2012) Fracture mechanics characterization of an anisotropic geomaterial. Eng Fract Mech 84:111–122
46. Zuo J, Wei X, Wang J et al (2018) Study on gradient failure mechanism and model of surrounding rock in deep roadway. J China Univ Min Technol 47(03):478–485
47. Zhang X, Zhang Q, Xiang W et al (2016) Analysis of partition fracture mechanism based on strain gradient theory. Chin J Rock Mech Eng (4):724–734
48. Wang M, Xie D, Li J et al (2013) Dynamic constitutive model of deep rock mass deformation and failure. Chin J Rock Mech Eng 32(6):1112–1120
49. Liu H, Xiao M, Chen J (2012) Parameterization study on spatial effect of excavation disturbance of complex underground cavern surrounding rock. J Sichuan Univ (Eng Sci Ed) 44(3):47–54
50. Chen ZH (2014) Stress distribution characteristics in rock surrounding heading face and its relationship with temporary supporting. Appl Mech Mater 568–570:1684–1689
51. Hou G, Niu X (2010) Ideal elasto-plastic solution of axisymmetric circular tunnel based on Levy-Mises constitutive relation and Hoek-Brown yield criterion. Chin J Rock Mech Eng 29(4):765–777
52. Ix C (1976) Fundamentals of rock mechanics. Blackwell Publishing, Blackwell, pp 132–166
53. Obert L, Duvall WI (1967) Rock mechanics and the design of structures in rock. Wiley, New York, 71–78
54. Cai M (2002) Rock mechanics and engineering. Science Press, Beijing, pp 21–41

Open Access This chapter is licensed under the terms of the Creative Commons Attribution-NonCommercial-NoDerivatives 4.0 International License (http://creativecommons.org/licenses/by-nc-nd/4.0/), which permits any noncommercial use, sharing, distribution and reproduction in any medium or format, as long as you give appropriate credit to the original author(s) and the source, provide a link to the Creative Commons license and indicate if you modified the licensed material. You do not have permission under this license to share adapted material derived from this chapter or parts of it.

The images or other third party material in this chapter are included in the chapter's Creative Commons license, unless indicated otherwise in a credit line to the material. If material is not included in the chapter's Creative Commons license and your intended use is not permitted by statutory regulation or exceeds the permitted use, you will need to obtain permission directly from the copyright holder.

The manufacturer's authorised representative in the EU is Springer Nature Customer Service Centre GmbH, Europaplatz 3, 69115 Heidelberg, Germany. If you have any concerns regarding our products, please contact ProductSafety@springernature.com

Printed and bound by CPI Group (UK) Ltd, Croydon, CR0 4YY

26/03/2026

02078941-0004